# ROADSIDE GEOLOGY

## of New Mexico

Halka Chronic

MOUNTAIN PRESS PUBLISHING COMPANY
MISSOULA 1987

**Library of Congress Cataloging-in-Publication Data**

Chronic, Halka.
   Roadside geology of New Mexico.

   Bibliography: p.
   Includes index.
   1. Geology—New Mexico—Guide-books.
   2. Roads—New Mexico
I. Title.
QE143.C47      1986      557.89      86-21748
ISBN 0-87842-209-9 (pbk.)


Mountain Press Publishing Company
P.O. Box 2399
Missoula, MT 59806
1-800-234-5308

# Dedication

To two old fuddy duddies and one kid cousin

# Contents

*The brilliant gypsum dunes of White Sands National Monument, majestic as New Mexico's summer clouds, sweep slowly northeastward before the push of the wind.* National Park Service photo.

# Preface

The "Land of Enchantment," New Mexico is a land of contrasts: arid deserts and forested highlands, salty playas and blowing dunes, rugged volcanic uplands and colorful plateaus, and a great river flowing unperturbed through a rift valley that quite literally slashes the state in two. New Mexico shows many faces to its residents and visitors—faces that in large part can be laid at the doorstep of the state's varied geology.

New Mexico lies at the crossroads of four geologic provinces. The Southern Rocky Mountains, the Colorado Plateau, the Great Plains, and the Basin and Range Province come together within the borders of the state, *their* borders as irregular as pieces of a jigsaw puzzle.

For purposes of this book, however, it seems appropriate to divide the state a little differently—but also in terms of geology and scenery. The Rio Grande Rift, the great sunken slice that bisects the state from Colorado to Texas, makes this easy: Divisions used throughout this book are The Western Highlands (chapter II), The Rift and the Rockies (chapter III), and East of the Rift (chapter IV). Since at its northern end the rift splits the southern tip of the Southern Rocky Mountains, they are included in chapter III as well. Chapter V describes the geology of national parks and monuments within the state.

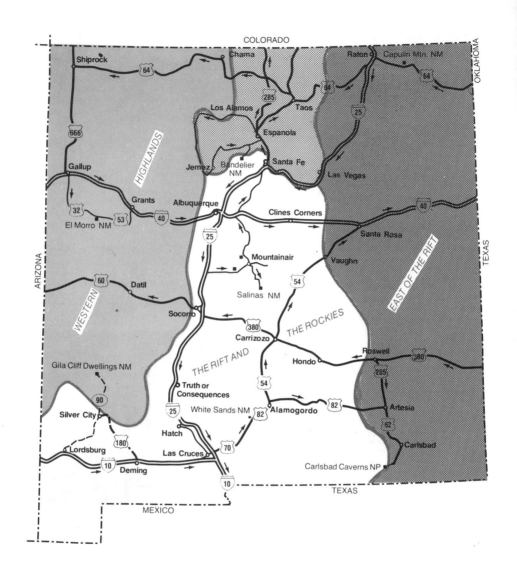

*New Mexico, cut in half by the Rio Grande Rift, reflects in its scenery and geology the three divisions used in this book.*

This book is designed especially, though not exclusively, for readers with little or no geologic training, for those curious about the world around them, eager for answers to apparent enigmas. Chapter I tells of geology in general, with emphasis on New Mexico's own geologic story. Each subsequent chapter starts with a rundown on the geology of one subdivision of the state, and should be read before traveling along any individual highway segment in that chapter. Road logs are arranged in numerical order within the chapters, interstate highways first, then US highways, and then state highways. A geologic map accompanies each unit.

Geologic terms are defined where first used, as well as in the glossary at the end of the book. Many geologic terms are adaptations of ordinary English words, and appear in standard dictionaries.

The book is oriented, obviously, toward travelers on New Mexico highways. However do be reasonable about its use. Read a section *before* going out on the highway, or have a passenger read as you go along. Neither the author nor the publisher is responsible for accidents or injuries resulting from use of this book.

New Mexico is about half desert, and the desert environment calls for extra care of yourself and your children. When you walk in the desert, watch where you put your hands and feet. Keep your car or a prominent landmark in sight as you walk. And stay clear of old mines—many are near collapse and some contain toxic gases. Hunting minerals and fossils is now prohibited or severely limited on public and Indian land.

Highways discussed in this book are logged in one direction across the state, with breaks at major cities and towns. Three loop trips, through regions of particular interest, connect with the main routes. For finding your way around, use any good road map; the one distributed by the New Mexico State Highway Department is fine, although some commercial maps do a better job of naming mountains, streams, small towns, and other features mentioned in the text. In the road logs, cultural and geographic names are used for orientation and location, but where named features are far apart or difficult to recognize, I've filled in with mileposts—those small green signs beside the road. Mileposts start with 0 at the western or southern state line, and read east or north across the state.

Highway speeds allow only broad geologic observations. So let me encourage you to stop often and to get out of your car to look more closely at the face of the land. Walk in the desert, hike in the mountains. In national parks and monuments, as well as in national forests, trails lead you close to many details that cannot be seen from your car window. On interstate highways, pause at rest stops for a good look at nearby rocks and surrounding views.

Examine rocks wherever you can: Handle them. Look at their color and texture, their durability or weakness. Ask yourself questions about them. What color are they? Are they rough or smooth? Can you see individual crystals or grains? Are they stratified (layered)? Do they contain fossils? Do you think they are in place, part of the rock mass of the Earth's crust, or have they been broken away and moved by water or gravity? How well are their components cemented together? Do their grains come loose in your hand? Can you identify their contacts with other rocks? Thanks to an arid climate, rocks in most of New Mexico are well exposed, easy to see and touch and feel.

Material in this book comes from published and unpublished sources and from field trip guidebooks of a number of professional organizations, notably the New Mexico Geological Society and the West Texas Geological Society, along with many of my own observations. Maps are derived from the Geologic Map of New Mexico published by the U.S. Geological Survey in 1965, updated by reference to the New Mexico Highway Geologic Map put out by the New Mexico Bureau of Mines and Mineral Resources in 1982. This excellent and inexpensive small map is available at many bookstores or from the New Mexico Bureau of Mines and Mineral Resources in Socorro. It includes a satellite mosaic of New Mexico, diagrams showing the succession of layered rocks in different parts of the state, and geologic sections across the state.

Geologic maps show the age of rock at the surface or just below thin soil and gravel layers. Usually they identify specific formations (recognizable rock units) or groups of formations. The same units are also shown in the geologic sections. Vertical scales are exaggerated in all the cross sections in this book—a practice that clarifies rock relationships but makes little hills look like big mountains and shallow valleys look like deep canyons. Exaggerating the vertical scale also steepens the slant, or dip, of tilted rock layers.

Symbols used for rock types are those commonly used to suggest the kinds of rock involved: dots for sandstone, a brick pattern for limestone, horizontal dashed lines for shale, which tends to split into thin slabs, and so forth. A legend for symbols and abbreviations used in this book is inside the front cover.

And finally I wish to thank those who have helped me prepare this book: the many geologists on whose work I relied for information, as well as David D. Alt, Donald W. Hyndman, David Gillette, and Felicie Williams, who read and critiqued the manuscript, and Edwin H. Colbert, Mike Williams, Frank E. Kottlowski, Ned Slater, and personnel of state and national parks and monuments, who offered many useful suggestions. I also thank NASA, the National Park Service, and the U.S. Geological Survey for contributing photographs. Where not otherwise indicated, the photographs are my own.

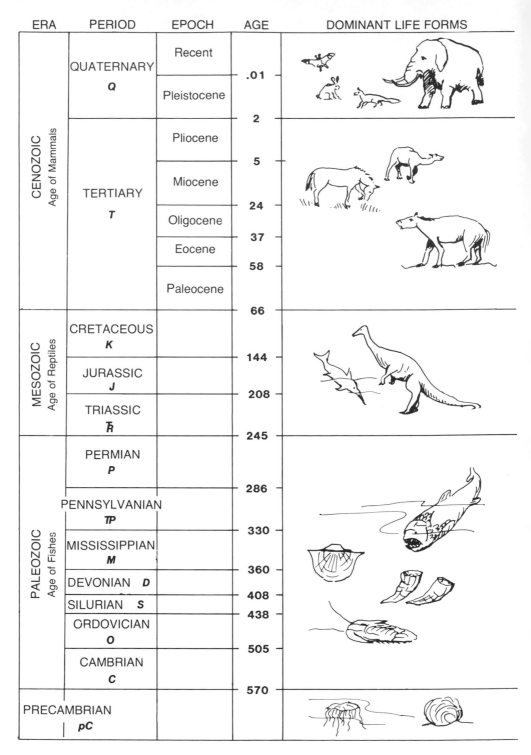

| ERA | PERIOD | EPOCH | AGE | DOMINANT LIFE FORMS |
|---|---|---|---|---|
| CENOZOIC Age of Mammals | QUATERNARY Q | Recent | | |
| | | Pleistocene | .01 | |
| | | | 2 | |
| | TERTIARY T | Pliocene | | |
| | | | 5 | |
| | | Miocene | | |
| | | | 24 | |
| | | Oligocene | | |
| | | | 37 | |
| | | Eocene | | |
| | | | 58 | |
| | | Paleocene | | |
| | | | 66 | |
| MESOZOIC Age of Reptiles | CRETACEOUS K | | | |
| | | | 144 | |
| | JURASSIC J | | | |
| | | | 208 | |
| | TRIASSIC ℞ | | | |
| | | | 245 | |
| PALEOZOIC Age of Fishes | PERMIAN P | | | |
| | | | 286 | |
| | PENNSYLVANIAN ℞P | | | |
| | | | 330 | |
| | MISSISSIPPIAN M | | | |
| | | | 360 | |
| | DEVONIAN D | | | |
| | | | 408 | |
| | SILURIAN S | | | |
| | | | 438 | |
| | ORDOVICIAN O | | | |
| | | | 505 | |
| | CAMBRIAN C | | | |
| | | | 570 | |
| PRECAMBRIAN pC | | | | |

*Geologic calendar*

| |
|---|
| Present erosion cycle trenches Pleistocene deposits, partly refills Rio Grande Rift valley. Basalt eruptions build cinder cones and lava flows near Grants, Carrizozo, and Capulin. |
| Cyclic erosion, product of repeated glacial cycles farther north, alternately trenches and fills the Rio Grande Valley. Small mountain glaciers develop in northern New Mexico mountains. Jemez volcano erupts and collapses. |
| Basins between ranges fill with debris eroded from surrounding mountains. Some drainage integrates: the Rio Grande becomes a through-flowing stream. |
| Increasing crustal tension creates basins and ranges of southern New Mexico. Intense volcanism builds and destroys many large volcanoes in the southwest part of the state. |
| The Rio Grande Rift begins to sink between two sets of faults. West of the still-sinking San Juan Basin, plateaus develop. |
| Debris from the Rocky Mountains fills the San Juan Basin. Mammals diversify, diversify, many the ancestors of modern forms. |
| Continued rise of Rocky Mountains and initial sinking of San Juan Basin accompanies westward drift of continent. Mammals flourish on land. Mineral-bearing intrusions form in parts of the state. |
| North America breaks away from Europe and starts to drift westward. Briefly, a vast sea covers parts of New Mexico. The Rocky Mountains rise to the north. Finally, a great extinction annihilates many forms of life, ending the Age of Reptiles. |
| Seas of sand sweep in wide deserts across northern New Mexico. Dinosaurs roam river floodplains and near-shore marshes. |
| Coastal plain, floodplain, and delta deposits spread across state, their sediments derived from ancestral Rockies. Explosive volcanism adds volcanic ash to these sediments. Dinosaurs appear. |
| Southern seas advance across much of New Mexico. A large barrier reef develops in the south, followed by drying up of the sea and creation of extensive salt and gypsum deposits. Locally, erosion removes some earlier sedimentary layers. |
| A southern sea covers much of New Mexico with sand, mud, and limestone. With the rise of the ancestral Rockies, sediments become coarser. |
| Widespread deposition of fossil-bearing marine limestone is followed by uplift and development of karst topography with solution caverns and sinks. |
| Marine deposits—limestone and shale—form in shallow seas. |
| Marine deposits form. Most are later eroded away. |
| Marine deposits—limestone and shale—form in shallow seas. The first fishes appear. |
| A western sea advances across the stripped Precambrian surface, depositing sandstone, shale and limestone. Shellfish are widespread and abundant: the Age of Fishes has begun. |
| Episodes of mountain-building and volcanism alternate with periods of marine and non-marine sedimentation. Intrusions of granite occurred roughly 1.35 billion years ago. Finally, a long period of erosion flattens the landscape. |

*The history of our island planet tells of continents that come together and drift apart. The Red Sea (upper arrow) and the Mozambique Channel (lower arrow) are widening rifts between continental masses. Another such rupture may be forming in New Mexico.* NASA photo.

# I
# Rocks and Time

Geology is the science of the Earth, the island planet on which we make our home. Classically, it concerns the crust of the Earth, the only part exposed to view. But within the last century geology has come to include data from well cores and from remote sensing techniques that explore the Earth's interior, the sea and sea floors, and the envelope of oxygen, nitrogen, and other gases that make up our atmosphere—appropriately the realm of atmospheric science but important to geologists because of its bearing on ancient climates and their expression in the rocks and the record of ancient life.

Despite remote sensors, despite computer models of our planet, despite geochemistry, seismology, geophysics, and other detailed analysis methods, data gathered in the field still contribute strongly to our understanding of the Earth. Geologists still hike and drive and fly to obtain first-hand knowledge directly from the Earth's surface. They study, define, measure, map, and identify detailed structures in individual rock exposures. They size up the larger structures of mountains and plains, and collect rock and fossil and mineral specimens for laboratory analysis. Field work is still the nucleus around which geologic knowledge is built.

*Our solar system was born in one of the spiraling arms of the Milky Way Galaxy, similar in many ways to the Andromeda Galaxy shown here.* © 1973 AURA, Inc., National Optical Astronomy Observatories.

## The Earth and its Origins

Our universe seems to have been born about eight billion years ago, in a supersized explosion fondly called the Big Bang. Matter—the stuff of stars, suns, moons, planets, and space dust and gas—exploded in all directions, shooting through the unconfined immensity of space and time. Blobs of matter occasionally collided, forming larger blobs. A collection of blobs, held together by gravity yet kept apart by their whirling motion, became our galaxy. And within the galaxy one particular blob, with a radiant ball of gases at its center, became our sun and solar system.

Gradually, as smaller and cooler blobs circling the hot central sun coagulated into balls of gas and solid material, the planets were born.

Planet Earth came into existence, as far as we can tell, about 4.6 billion years ago when a tight ball of cooling gas condensed and developed a thin but solid crust around a hot, partly molten interior. No traces of this ancient crust remain. The oldest rocks now known, nearly four billion years old, are hand-me-downs reformed from older rock.

As the interior of the young planet cooled, it separated, as water and oil separate, into heavier and lighter components, the Earth's core and mantle. The core, 4400 miles in diameter, is composed of nickel and iron. The mantle around it, 1800 miles thick, is partly solid but has a liquid (or at least semiliquid or plastic) layer that has a great deal to do with the behavior of the Earth's crust, as we shall see. Then, outside the mantle, floating on its semiliquid layer, is the crust, three miles thick under the oceans, 20-25 miles thick under the continents.

The Earth's crust moves in unison with the upper, or outer, part of the mantle, a pairing to which geologists now give the name lithosphere. Partly mantle, partly crust, the lithosphere

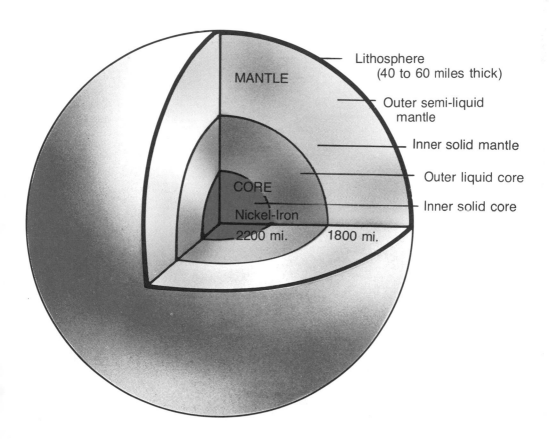

seems to be about 40 miles thick under the oceans and between 60 and 200 miles thick below the continents. Relative to the size of the Earth, that is pretty thin, hardly more than a film, a fairly pliable yet easily broken film that can be rumpled or torn or moved about by stirrings in the more fluid part of the mantle below.

Heat generated in the mantle by decay of radioactive minerals runs the Earth's stove. It causes the less dense semiliquid part of the mantle to rise by convective heating, and to roll over and, growing cooler, to plunge downward again, as thick soup churns and rolls in a pot on the stove. The churning movement of the mantle is extremely slow by human standards. But as the mantle moves, the thin film of the lithosphere is carried along, pushed up where the boiling lifts it, moved horizontally at the top of the roll, and pulled downward as the cooling mantle descends once more toward its source of heat. Movements of the lithosphere create mountains and continents, oceans and seas, and are back of most of the geography and geology we see today.

## Wandering Continents, Widening Seas

According to the Theory of Plate Tectonics, accepted now by nearly all geologists, movement in the upper mantle has broken the lithosphere, producing twelve large, relatively rigid plates, somewhat like the plates of a turtle's shell, interspersed with several smaller ones. The plates of the lithosphere are made of two sorts of rock—continental rock, which by and large is light in color and forms the continents, and dark oceanic rock, which makes up the floors of the major oceans. Most plates—particularly the large ones—are partly oceanic, partly continental. The oceanic part, though thinner, is denser than the continental part.

Plates are not static: During the long history of the Earth they have moved about and jostled one another, combined and broken apart, sunk or risen or partly slid over one another, because of the convective churning of the mantle. We are reminded of this slow roll by earthquakes and volcanic eruptions and even by the gradual drowning of some of the works of man.

Oceanic parts of the large plates are bordered by mid-ocean ridges or by deep oceanic trenches that lie near the edges of some continents and oceanic islands. At mid-ocean ridges, two

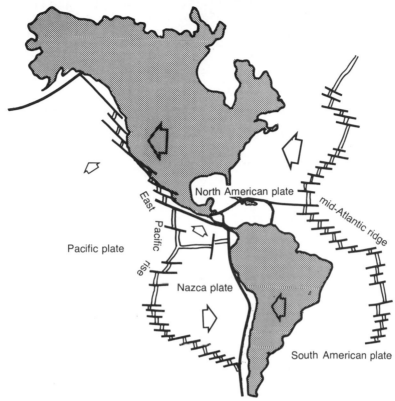

*With formation of new crust along the mid-Atlantic Ridge, the North American Plate has moved westward (arrow) across the East Pacific Plate, now mostly hidden under western North America. Similarly, the South American Plate is overriding the Nazca Plate. Mid-ocean ridges are offset by numerous transform faults.*

adjacent plates slowly move apart as molten rock wells up from the mantle to form new oceanic crust, which then becomes part of the plates on either side. At trenches, the continental part of one plate overrides the oceanic part of another, which then slowly plunges downward and reheats, once more becoming part of the mantle. Where two continental parts of plates collide, great mountain ranges rise, the crumples on the Earth's bruised surface.

The plate we're concerned with in this book is the North American plate, stretching from the mid-Atlantic Ridge to the Pacific Coast. This plate is drifting southwestward as the Atlantic Ocean widens, a process that has been going on for 100 million years. In New Mexico, two parts of the North American plate seem to be very slowly separating, breaking apart along

*Joints caused by earth movements, cooling of igneous rocks or release of pressure show up in practically all rocks. Columnar joints in the Cimarron Palisades, shown here, are due to slow cooling of a flat sheet of molten rock well below the Earth's surface. Fine horizontal joints developed as overburden was removed by erosion.*
E.F. Patterson photo courtesy of U.S. Geological Survey.

deep, parallel faults, leaving a slender down-dropped zone that is the Rio Grande Rift. We'll come back again and again to this slender split in the Earth's crust in the pages that follow.

## Folds and Faults

As the plates drift about, pushed up and pulled down, tearing apart, crunching into, or sideswiping each other, the rocks of which they are made are subjected to all kinds of stresses. They react by bending and breaking. It's almost impossible to find a rock of any size that doesn't show these signs of stress. Fractures, called joints, mark almost all rock outcrops and serve as avenues for weathering and erosion. Folds, where layers of rocks bend, are common features. Faults—fractures where measurable movement has taken place—are also common, some with thousands of feet or even miles of movement.

We classify faults by the angle of inclination of fault fracture surfaces and by the relative movement of the two sides. Geologists recognize several types of folds and faults, shown on

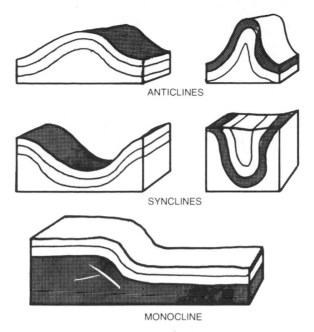

ANTICLINES

SYNCLINES

MONOCLINE

*Three kinds of folds are easily recognized when they occur in layered (stratified) sedimentary or volcanic rocks.*

accompanying diagrams. In normal and reverse faults the plane of the fault is steep, whereas in thrust and extension faults it is shallow—often nearly horizontal. Listric normal faults, which come in bunches, may be nearly vertical at the surface, but they curve with depth, joining other faults in their bunch to form single nearly horizontal extension faults.

Note that normal, extension, and listric faults are all caused by horizontal tension or pulling apart in the Earth's crust. Reverse and thrust faults, on the other hand, are caused by horizontal compression or pushing together. But the movement in both cases is relative: Both sides may actually move in the same direction, one farther than the other.

*Tilted sedimentary rocks weather into hogback ridges, the more resistant layers capping individual hogbacks.*

7

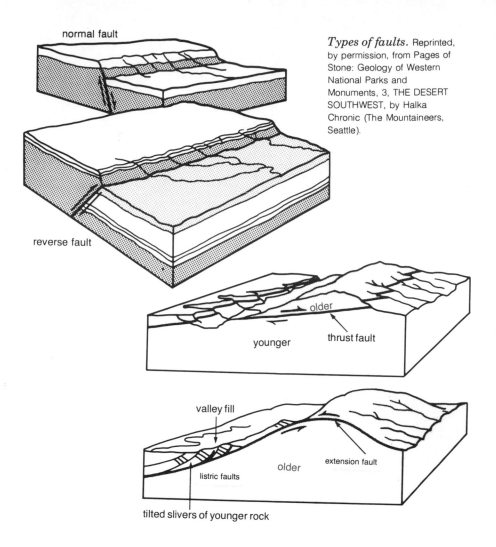

normal fault

reverse fault

older

younger

thrust fault

valley fill

older

extension fault

listric faults

tilted slivers of younger rock

*Types of faults.* Reprinted, by permission, from Pages of Stone: Geology of Western National Parks and Monuments, 3, THE DESERT SOUTHWEST, by Halka Chronic (The Mountaineers, Seattle).

It's important to remember that faults are rarely as simple as those shown in the diagrams; numerous small, individual faults commonly make up large, complex fault zones.

Let's get back to the Rio Grande Rift. The two more or less parallel fault zones that cross New Mexico from north to south, bordering the rift, came into being about 30 million years ago, and movement along them is still going on today. As tension pulled the two halves of New Mexico apart, the central north-south sliver between the fault zones dropped downward, in places as much as 30,000 feet. Rift-bordering volcanos spewed

*Two great rift valleys cut southwestern United States. Bordered by major faults that reach down to the mantle, they suggest that the continent is slowly splitting apart.*

dark basalt derived from the Earth's mantle, telling us that the faults reach all the way through the crust, forming avenues that allow mantle material to rise to the surface. The major pulling apart of the crust that is responsible for the rift may be due to upward "boiling" of mantle material like that along mid-ocean ridges. Possibly this pulling apart is the beginning of a new narrow seaway similar to the Red Sea or the Gulf of California. And the Rio Grande Rift may eventually—millions of years from now—widen to form a new ocean basin comparable in its size and origin to the Atlantic Ocean.

## Uplift and Erosion

As plate movement, folding, and faulting work to build up mountains and create differences in elevation over broad regions, water, wind, and gravity combine to tear them down, to remove the rock materials of which they are made, to flatten the land. Water, wind, and gravity play triple roles, wearing away, transporting, and redepositing rocks and rock debris.

Uplift feeds erosion: The higher an area is lifted, the greater are the opportunities for wearing away the rock. Steep-coursed streams, well laden with boulders, gouge mountain canyons, whereas slow-moving rivers have little cutting power and indeed may deposit more than they erode.

Rock itself is a chief tool of erosion—rock whirled and bounced and hammered into other rocks, gradually breaking itself and the rock it strikes into smaller and smaller pieces, finally into sand and silt and clay. Ice and frost play parts as

9

*A large volcanic neck, made up of hard igneous rock that cooled and solidified in a volcano's conduit, stands out above the silty desert of northwestern New Mexico.*

well, particularly near mountain summits, where freezing is a nightly affair. Water seeping into joints and cracks expands as it freezes, and gradually splinters the rock, making it easy prey for fast-running rivulets or for gravity, which by itself brings down landslides, rockslides, and tumbling blocks of rock.

Some rocks are better able to resist erosion than others, and remain as mountains or hills while less resistant rock is worn away. Magma that cooled and hardened slowly in a volcano's conduit, for instance, may remain as a prominent part of an otherwise leveled landscape—a volcanic neck. Fairly resistant rocks like sandstone and limestone may cap flat-topped mesas and plateaus or tilted cuestas and hogbacks as weaker mudstone and siltstone wear away.

Slower and less obvious forces contribute to erosion too—the gradual weathering of rock as it is attacked by rain, snow, and other atmospheric agents, solution and chemical alteration by products of plant and animal metabolism, and the gradual solution of limestone by groundwater fortified by atmospheric and soil acids. Even the burrowing of small animals and the hoof-marked trails of larger ones hasten erosion.

*New Mexico is a land of mesas, most of them capped with resistant sandstone (as here), lava flows, or caliche-impregnated gravel.*

10

HOGBACK

BUTTE

resistant
caprock
layer

CUESTA

MESA

Landforms shaped by water begin with erosion. Canyons and cliffs, sharpened peaks and ridges, caverns and sinkholes are common throughout New Mexico. Landforms resulting from deposition are common as well. In an arid climate, where plant cover is sparse, occasional heavy rains remove great quantities of unprotected rock material from normally dry canyons, sweeping it valleyward and depositing it in steep alluvial fans that radiate from canyon mouths. Mudflows add to fan thickness. Most of the water from mountain torrents sinks into the coarse gravel of the alluvial fans, natural reservoirs for water. But after real "cloudbursts," water may sheet off the fans into adjacent rivers or onto flat valley floors, in the latter case commonly filling shallow undrained playa lakes. Both alluvial fans and playa lakes are important in gradual filling in of basins between mountain ranges.

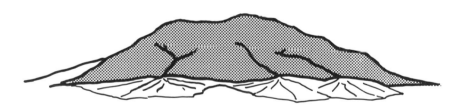

*Alluvial fans (top) develop at the mouths of mountain canyons. Eventually they may coalesce, as shown here, to form an alluvial apron or bajada. The shimmering playas (bottom) of undrained intermountain basins are among the flattest surfaces on Earth.*

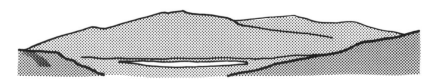

11

Flowing rivers are of course rare in desert regions, except for those nurtured by flow from high mountains beyond the limits of the desert. The Rio Grande, largest river in New Mexico, rises in the San Juan Mountains of Colorado, as does the San Juan River in the northwestern part of the state. The Canadian and Pecos rivers originate in the Sangre de Cristo Mountains of northern New Mexico. Many of the rivers shown on roadmaps are dry a good part of the year: the Rio Puerco, the Rio Salado, the Rio Hondo, the Conchas, and others. Some with natural but seasonal flow are now dry year-round because dams hold back spring runoff.

There is plenty of evidence in New Mexico showing that water—running and otherwise—once played a larger part in development of the landscape. Broad gravelly terraces that stairstep the edges of many valleys were once floodplains in their own right, then were cut into by the very rivers that created them. The highest terraces are the oldest, as the drawing shows. Wave-cut benches in undrained valleys show that lakes once existed where today there are only salt flats—evidence of a wetter climate in the past.

In desert regions, wind plays a role in erosion, too. Basins deepen as wind scours away fine sand and silt. Small hollowed blowouts, and rocks pitted and undermined by wind-driven sand, testify to that role, as does the peculiar pebble-covered surface called desert pavement, where wind has removed sand, silt, and clay particles, leaving a surface of tightly packed pebbles. The ability of wind to sweep up and remove quantities

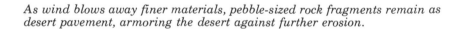

*As wind blows away finer materials, pebble-sized rock fragments remain as desert pavement, armoring the desert against further erosion.*

12

of fine particles from the desert surface is all too evident during dust storms: A single large dust storm can make off with millions of tons of dust. On a smaller scale, the whirling winds of desert dust devils lift and carry sizeable sand particles— even gravel—for shorter distances. Plowing and cultivation, particularly coupled with drought, play their part in setting the stage for wind erosion.

Wind often carries silt and clay particles far from their original sites. Sand grains are bounced along the surface until the wind's energy is dissipated or turned aside. Dunes, piled accumulations of wind-blown sand, occur in many parts of New Mexico.

## Rocks and Minerals

In addition to their classification into oceanic and continental, rocks, in general, can be divided by their mode of origin into three great categories: igneous, sedimentary, and metamorphic rock.

*Most igneous rocks resist erosion well. Here Shiprock, a volcanic neck, and a crack-filling dike tower over the surrounding landscape of softer rock.*

Igneous rocks originate from molten rock, magma, that comes from sources deep within or below the Earth's solid crust. If magma pushes through to the surface and cools rapidly, it forms volcanic rock. Such rock solidifies so fast that mineral crystals have little or no time to form, so volcanic rocks are normally quite fine-grained, with crystals that are invisible to the naked eye. Scattered visible crystals grew early, at depth, with slower cooling. Common volcanic rocks include lava, tuff formed from volcanic ash, and volcanic breccia, a broken and recemented mixture of lava and tuff.

Tuff comes in two varieties: ashfall tuff, which settles from the great mushroom clouds of volcanic ash emitted by some erupting volcanoes, and ashflow tuff, which results when very hot volcanic ash shoots down a volcano's flank and is still hot enough when it comes to rest to fuse into a porcelain-like mass.

Volcanic rocks can also be classified by their chemical or mineral composition. Those of basaltic composition are dark with iron and magnesium minerals. Those of silicic composition are lighter in color and contain large amounts of silica in minerals such as quartz and feldspar. It is clear that basaltic magma comes from melting in the Earth's mantle, whereas silicic magma comes from melting of continental crustal rocks at shallower levels.

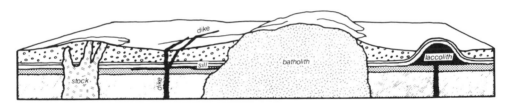

*Shapes and sizes of* intrusions *vary. They are named according to size and their position relative to stratified rocks.* Intrusive rocks *harden below the surface; those shown at the surface here have been bared by erosion.*

Magma that doesn't reach the surface, but cools and hardens within the crust, becomes intrusive igneous rock. Because it cools slowly, over hundreds, thousands, or millions of years, there is ample time for visible crystals to develop. Granite— light-colored and coarse-grained—is a common intrusive rock. As granite cools and shrinks, it often cracks, and still-fluid constituents seep into the cracks, there to crystallize into particularly coarse-grained pegmatite veins.

Some intrusive rocks contain scattered large crystals set in a

14

*In granite, crystals are all about the same size.*

*Porphyry contains large crystals scattered in a finer matrix.*

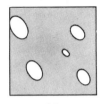

*Crystals in volcanic rocks are often invisible without magnification. Gas bubble holes are common.*

*Textures of igneous rocks*

fine-grained matrix. We call these rocks porphyry. Porphyries are important in New Mexico in that they are often the sites of enrichment by copper-, silver-, and gold-bearing minerals.

Sedimentary rocks originate, as their name implies, from sediment—remains of older rocks or animal shells broken up and redeposited. These materials may be deposited by water in river valleys, lakes, or seas, or by wind as sand dunes or layers of dust. Glacial ice also deposits sedimentary rocks, but ice age glaciers hardly touched New Mexico, and there are few glacial deposits near the highways described in this book.

Sedimentary rocks are classified by the size of the particles of which they are composed: Conglomerate starts out as gravel, and contains stream-rounded pebbles, boulders, and sand cemented together by calcium carbonate or some form of silica. Sandstone, siltstone, mudstone, and claystone are made of—you guessed it!—sand, silt, mud, and clay. Mudstone and claystone frequently break easily along thin partings, in which case they may be called shale. See the accompanying chart for grain sizes in these different rock types. Limestone usually forms from accumulations of lime mud or animal and plant shells.

Sedimentary rocks are nearly always stratified, arranged in

*Conglomerate, formed from gravel, contains both large and small rock fragments rounded by bouncing and grinding against one another in running water. The matrix is usually sand.*

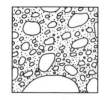

*In breccia, rock fragments are angular. Matrix may be clay, finely crystalline rock or minerals which grew later around the fragments, cementing them together.*

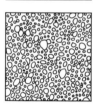

*Sand is made up of particles 1/32 to 1/4 inch in diameter. Grains are usually rounded.*

*Siltstone particles are less than 1/32 inch in diameter.*

*Textures of sedimentary rocks*

*Sedimentary rocks are layered or stratified. Here, layers of limestone are separated by thin beds of weaker siltstone. Nearly always, the layers were deposited in horizontal position.*

16

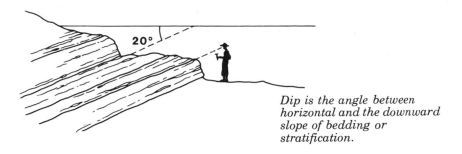

*Dip is the angle between horizontal and the downward slope of bedding or stratification.*

strata or layers. Individual layers are for some reason called beds. Most sediments are deposited in lakes, oceans, and river floodplains and deltas in horizontal layers which later, when they have turned to rock, may be bent or tilted or broken by movements in the crust—a feature useful to geologists trying to decipher the histories of particular regions or features. When once-horizontal layers are tilted, we speak of dip, the angle of tilt below horizontal.

Sedimentary rocks often contain fossils, the remains, impressions, or trails of animals and plants that lived and died at the time the sediments were deposited. They may also be marked with mudcracks, ripple marks, or salt crystal impressions, useful indicators of the environment in which the rocks formed. Some sedimentary rock is crossbedded, with sloping laminae that show it was deposited by flowing water or wind.

Metamorphic rocks also start out as other rocks; they are altered by heat and pressure rather than by being broken up and redeposited. Most alteration, or metamorphism, occurs deep below the surface, beneath a weighty overburden of other rock. The two most common metamorphic rocks are gneiss, a massive, banded, but otherwise granite-like rock, and schist, a rock with a pronounced tendency to break into thin parallel flakes.

That brings us to minerals, the building blocks of which rocks are made. Minerals each have definite physical and chemical properties; often they display clearcut colors and characteristic ways of crystallizing—all factors that help in their identification. New Mexico has its share of unusual and beautiful minerals, and its share of rockhounds who collect them. There are numerous rock shops in the state, as well as good museum mineral collections—notably those at the New

Mexico Institute of Mines and Mineral Resources in Socorro and the New Mexico Museum of Natural History in Albuquerque. Many of the minerals displayed in such collections are exceptionally large specimens found during mining operations. It is fun to search old mine dumps for smaller versions. And New Mexico has one state park that encourages mineral collecting!

Some of the less flamboyant but common rock-forming minerals are listed below:

Quartz is a glass-clear or milk-white mineral so hard that it can't be scratched with a knife. Like glass, it breaks along curved surfaces. Some attractive New Mexico varieties, colored by minute amounts of other minerals, are pink rose quartz, lavender amethyst, and gray smoky quartz. Quartz is common as glassy-looking grains in granite and metamorphic rocks, and makes up most sand and sandstone.

Feldspar is a family name for a group of translucent pink, gray, or white minerals that may be a little harder than a knife, but that can be distinguished from quartz because they normally break along flat faces that reflect sunlight. Feldspar crystals are the most abundant components of granite and many other igneous and metamorphic rocks.

Micas are black or silvery minerals that separate into shiny, flat, paper-thin sheets. Mica can be scratched easily with a knife or even with a copper penny. Black biotite and white muscovite are common in granite and other intrusive rocks as well as in schist, a metamorphic rock.

Calcite is a white or light gray mineral that makes up limestone, but that also occurs as well developed crystals. It can be scratched with a knife but not with a fingernail. Some varieties are transparent. Geologists drop a little dilute acid on rock samples to see if they contain calcite; if they do, the acid fizzes. Calcite is also soluble in natural weak acids, such as rainwater that has filtered down through soil layers, an attribute that leads to the development of limestone solution caverns.

Hematite is an iron oxide easily recognized by its rust red color. Small amounts of hematite lend pale pink to deep red tones to many of New Mexico's sedimentary rocks.

Limonite is a dull, mustard yellow iron oxide that also colors other rocks, giving them tan or pale yellow hues.

18

Amphiboles appear as dark green to black, rodlike crystals in igneous and metamorphic rock.

Olivine, an olive green or yellowish mineral, occurs as small, roundish crystals dotting some basalt.

Gypsum, quite common in New Mexico, is a translucent, easily scratched (with a fingernail), fairly soft, white or light gray mineral formed when sea water or salt lakes dry up. In the red, hematite-tinted sedimentary rocks of northern New Mexico, gypsum commonly appears as thick gray-white layers or as thin white veins and veinlets. Gypsum is somewhat soluble, and its gradual solution by groundwater is reponsible for interesting collapse features in eastern New Mexico.

Salt, also formed as sea or lake water dries up, is common in New Mexico, too. Like gypsum, it may be dissolved and removed by groundwater, causing collapse of overlying rocks.

## Volcanoes

Volcanoes form where molten magma reaches the Earth's surface. They vary in size from small spatter cones to immense

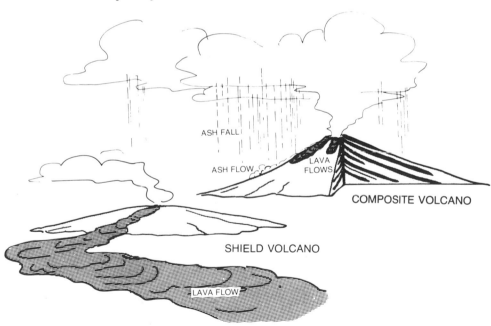

*Thick silicic lava builds tall, graceful stratovolcanoes in which stubby lava flows alternate with layers of volcanic ash. Basaltic lava, on the other hand, flows easily and forms low shield volcanoes or spreads out across the landscape in flat sheets.*

*Sierra Grande in northeastern New Mexico is a shield volcano.*

mountains. The shape and eruptive behavior of volcanoes depends on the chemistry of the rocks of which they are built.

Normally, basalt magma is runny and flows out fairly quietly as thin, wide-spreading lava flows. Where flows do pile up, they form broad-based shield volcanoes, of which there are several in New Mexico. If basalt magmas take on a lot of water, they may boil and foam upward in their conduit and throw out foamy bits of lava known as cinders, building small volcanoes known as cinder cones. Or, if there is not as much gas, the hardening lava may contain lots of little bubble holes or vesicles, some of which may later fill in with opal or other minerals carried in solution by groundwater.

Silicic magma is so stiff and pasty that it tends to erupt as lava domes or as short, thick lava flows. But silicic lava often contains much dissolved water, setting the stage for explosive eruptions like that of Mt. St. Helens in 1980, which spread fine volcanic ash all over the Northwest. So in silicic volcanoes the short, stubby flows frequently alternate with volcanic ash to build graceful composite volcanoes like Fujiyama, Mt. Vesuvius, and Mt. Hood. As we'll see, the Mt. St. Helens eruption was a mere baby; long-ago New Mexico eruptions spewed a hundred times as much volcanic ash. Some of the ash rose in giant mushroom clouds, to be carried around the world by stratospheric winds. And some, still incandescent, propelled by its own expanding gases, sped down volcanic slopes as ashflows that came to rest still so hot that they fused into welded or ashflow tuff. In the end, so much volcanic ash was expelled that many of the volcanoes collapsed into the partly-emptied magma chambers below them, leaving large, circular, cliff-walled depressions known as calderas to mark where they had once stood.

## Geologic Time

Time, for geologists, goes back about 4.6 billion years, to the birth of the Earth. The oldest rocks now known were formed about four billion years ago. The oldest rocks in New Mexico are almost two billion years old.

Geologists only decades ago learned how to measure the age of rocks with a fair degree of accuracy. This they do by radiometric dating, which measures the decay of radioactive elements, believed to take place at steady rates. They can also now determine the age of some rocks by relating their natural rock magnetism to the history of magnetic reversals in the Earth's magnetic field, when north and south magnetic poles switched places—a process based on previous radiometric dating. Both methods are reasonably precise, and give ages in millions or, for the youngest rocks, thousands of years.

*These typical Cretaceous marine fossils come from the Dakota Sandstone and Mancos Shale of New Mexico.* W.A. Cobban photo courtesy of U.S. Geological Survey.

Earlier geologists relied on a time scale developed through years of careful study, a geologic calendar by which they could relate the age of one rock to that of another. The calendar relied on two principles that govern all of geology: 1) In undisturbed sedimentary rock layers, those on the bottom are the oldest, those on top are the youngest. And 2) when one rock cuts across another, the one doing the cutting is younger than the one that is cut. Using these two principles, geologists worked out a fairly accurate sequence for many areas. And to pin their sequences together so that they could compare rocks of one area—even one continent—with another, they had fossils.

Early geologists discovered that certain sedimentary rock layers contain certain kinds of fossils or groups of fossils. And once it became clear—partly from the study of fossils—that life on Earth was not static, but had evolved over time, fossils were recognized as time-markers. Queer little corkscrew-shaped snail shells in one layer, skinny clam shells in another, odd-shaped sea urchin spines, perhaps, or assemblages of certain crinoids (sea lilies) and brachiopods (lamp shells), and corals, signaled rocks of a certain age—whether they occurred in England or Russia or in colonies across the sea.

Using these time markers, geologists began to put together a worldwide calendar. The calendar divides geologic time into eras, the largest units, and subdivides eras into periods and periods into epochs. The eras are named Paleozoic (old life), Mesozoic (middle life), and Cenozoic (recent life). The name Precambrian is given to the oldest rocks, once thought to have formed before there was life on Earth. We now know that living things flourished in at least late Precambrian time, but did not grow hard, readily preserved shells.

Names of Paleozoic periods were derived mainly from European places and people: Cambrian from Cambria or Wales, Ordovician and Silurian from tribes that inhabited Wales in Roman times, Devonian from Devon in England, Permian from Perm in Russia. Mississippian and Pennsylvanian are the North American equivalent of Europe's Carboniferous (coal-bearing) period.

Mesozoic period names are also from Europe: Triassic because German Triassic rocks subdivide into three main groups, Jurassic for the Jura Mountains of eastern France, and Cretaceous in honor of white chalk cliffs near Dover and along the coast of France.

Cenozoic period names are relics of an early classification. Tertiary rocks (note that we can say "Tertiary rocks" as well as "Tertiary time") were formed before the onset of the great Ice Ages, and Quaternary rocks were formed during and after the Ice Ages, in the last two million years. Some Quaternary "rocks" are not even rocks yet—stream gravels, dunes, beaches, that sort of thing. Others, particularly volcanic rocks, are quite hard.

Epoch names differ from continent to continent, or even from different parts of the same continent. In this book, epoch names are used only for the Cenozoic Era.

Other types of geologic divisions were also useful in the early days of geologic research, and they, too, still remain in use today. Recognizable, mappable units of rock, usually separated from other mappable units by visible differences in rock types, are called formations. Formations, which in some cases are combined into groups, are often separated by unconformities that represent times of erosion or gaps in deposition. Such gaps can be established by mismatches in the attitude (tilt or dip) of individual rock layers, identification of old erosional surfaces, the presence of conglomerate at the bottom of a layer, old soil layers, or gaps in sequences of fossils.

Radiometric dating now gives us more exact boundaries between the time units. It leans on the premise that radioactive minerals decay at steady rates. Uranium, for instance, steadily breaks down into helium and lead. Uranium occurs in many igneous rocks that didn't contain, to begin with, measurable amounts of lead of a particular isotope or atomic structure. So by analyzing igneous rock, measuring its relative amounts of uranium and lead, having particular atomic structures, we can determine just how long ago those igneous rocks formed. Similarly, radioactive potassium decays into calcium and argon—but in this case it is the argon that is measured. And rubidium decays into strontium. Carbon 14, another useful tool, forms from nitrogen in the atmosphere and is present in all living things; it gradually reconverts into nitrogen. Measurements of carbon 14 in wood, shells, and other organic matter can be used to date that matter and the rock containing it back to about

40,000 years. A handy tool for archeologists, too, as everything from Folsom Man's campfire charcoal to the roofbeams in a New Mexico pueblo can be dated this way.

Radiometric dates have now been incorporated in the earlier, completely dateless calendar. The dates undergo frequent revision and refinement, giving us more and more exact knowledge about the geologic past. Dates used on the geologic time scale in this book are those adopted by the Geological Society of America in 1983.

## Mines and Mining

Ever since the arrival of the Spanish in the 16th century, geology has played an important role in New Mexico's economy. Even before that, Indians gathered and mined chert and obsidian for spearpoints and arrowheads, clays for pottery and body decorations, turquoise for ornamentation and ceremony, coal for fuel, and salt for cooking and trade.

Early Spanish attempts to find gold and silver in New Mexico ended in failure: The legendary Seven Cities of Gold sought by Marcos de Niza and Vasquez de Coronado turned out to be mud villages. But later Spanish settlers mined lead and turquoise near Cerrillos (south of Santa Fe) and in 1798 discovered copper at Santa Rita in southwestern New Mexico— where it is still being mined.

In 1828, 21 years before the California gold rush, gold was discovered in the Ortiz Mountains south of Santa Fe. Though plagued by Indian attacks and a perennial shortage of water,

*Southwestern New Mexico's mountains are dotted with old copper mines. This one, photographed in 1916, was north of Silver City near the town of Mogollon.* H.G. Ferguson photo courtesy of U.S. Geological Survey.

24

*A turn-of-the-century coal mine near Raton supplied railway and local needs as well as those of out-of-state users.* W.T. Lee photo courtesy of U.S. Geological Survey.

placer miners there managed to turn out several millions of dollars' worth of gold. In placer mines, gold-containing stream gravels were dug up and washed in gold pans and riffle-bottomed sluices, until the heavy gold settled out from the lighter gravels—a laborious, water-consuming process.

Copper mining didn't come into its own in New Mexico until late in the 19th century. As Indian raids were subdued and then eliminated, old Spanish mines near Silver City were reopened and new deposits were found in the same general area. Lead was mined near Las Cruces, and placer gold in many other localities, most of them along the valley of the Rio Grande or in the Sangre de Cristo Mountains. Gold-bearing veins, in contrast to placer gold, were found by tracing placer deposits upstream. Coal was discovered in several areas, some of them conveniently close to gold and copper mines.

Silver was discovered at Magdalena, west of Socorro, in 1863, and at Georgetown in Grant County a few years later. Then huge orebodies of lead, also near Magdalena. All over the state (then a territory) new discoveries were made and new mines opened up: at Socorro, Red River Canyon, Hondo Canyon, Chloride, Elizabethtown, Twining, Pinos Altos, Ute Creek, Cerrillos, White Oaks, Silver City, and other sites.

Most of these old New Mexico mines are silent now, victims

25

of depleted ore bodies and rising costs. Those that are operating have gone over to open-pit mining, made economical by the use of giant earth-moving machinery. Coal is mined now in the plateau country of northwest New Mexico, and copper-silver-gold ores are still produced in the Silver City area. Manganese, zinc, and lead come from the Silver City region, too. Large molybdenum mines operate in the Sangre de Cristo Mountains east of Questa. Limestone, gypsum, potash, and salt are produced from Pennsylvanian and Permian sedimentary rocks that underlie the eastern half of the state. Uranium is actively mined near Grants, though not as strenuously as during the uranium booms of the 1950s and 1970s.

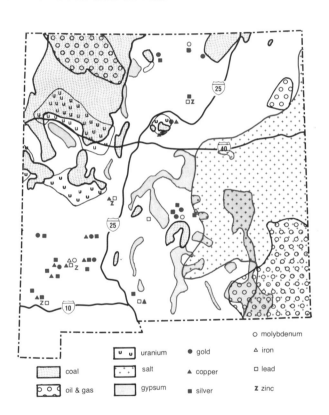

*Northwestern New Mexico has rich reserves of oil, gas, coal, and uranium – the energy minerals. In the southeast are more gas and oil, as well as widespread deposits of salt and gypsum. The belt between is rich in gold, silver, copper, and other minerals.*

## Oil and Gas

Other kinds of mineral wealth—petroleum and natural gas—come from the San Juan Basin in the northwestern corner of New Mexico, and from the Permian Basin in the southeastern corner of the state. Both of these large geologic basins are broad downwarps filled in with thick sequences of sedimentary rocks containing both source rocks and natural reservoirs for oil and gas, substances produced by slow decomposition of plant and animal material within the source rocks. Reservoir rocks of the San Juan Basin are Cretaceous and Tertiary sandstones; those in the Permian Basin are Permian limestones riddled with caverns and passages of all sizes.

## New Mexico's Geologic History

New Mexico's history is eventful—mountain-building, invasions and retreats of seas from east and west, development of large deltas and broad river floodplains, advancing deserts, exploding volcanoes, and pronounced episodes of faulting. But the record of these events is fragmentary—a patch of granite here, dune-deposited sandstone there, fossil-rich limestone or the ring-shaped faults of an old caldera somewhere else. Putting successive geologic events into their proper chronologic order, geologists working in the state have defined the sequence of events outlined below.

Of New Mexico's Precambrian history we know relatively little. Precambrian rocks are exposed in mountain ranges scattered over the western two thirds of the state. These rocks—gneiss, schist, and partly metamorphosed sedimentary rocks—tell us that two billion years ago ancient New Mexico,

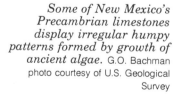

*Some of New Mexico's Precambrian limestones display irregular humpy patterns formed by growth of ancient algae.* G.O. Bachman photo courtesy of U.S. Geological Survey

27

swept by the sea, was also the site of volcanoes and other mountains. At times huge volcanic explosions deluged the land with volcanic ash, in a scenario repeated several times in New Mexico's past.

Around 1.35 billion years ago, after millions of years of erosion, only the roots of the former mountains remained. They were intruded, in a new round of mountain-building, by great masses of magma that slowly cooled into pinkish or reddish granite crisscrossed with coarse-grained pegmatite veins. Then once more the land was worn down, the mountains planed away. For millions of years it lay almost featureless, sloping gently toward a distant western sea.

During part of the Paleozoic Era, North America was part of a megacontinent that geologists call Pangaea, which also included Europe, Asia, Africa, and Antarctica. Around 570 million years ago, as the western part of this megacontinent tilted just a little, a sea crept across it, depositing a succession of marine sedimentary rocks—sandstone, siltstone, and limestone. Sedimentary rocks rich in marine fossils alternated with continental deposits that accumulated on river floodplains and deltas, or in vast deserts of dunes that marched across the land—another theme repeated several times in the history of the state. During parts of the era no deposits formed, or if they formed were later removed by erosion.

Late in Paleozoic time, during the Pennsylvanian and Permian periods, mountains rose in north-central New Mexico, part of the ancestral Rocky Mountains that stretched north into Colorado more or less along the line of the present Rockies. Far south of the mountains, in marine embayments along the present New Mexico-Texas border, a great barrier reef developed. As this and other embayments were cut off from the sea, their waters evaporated, leaving vast deposits of salt, gypsum, and potash.

During Mesozoic time, continental sediments were deposited around the new mountains, gravel and sand of alluvial fans, broad sheets of sand and silt of floodplains and deltas. These sediments now furnish the colorful red ledges and pink cliffs seen in the northern part of the state. Dinosaurs roamed the Mesozoic world, and violently erupting volcanoes added volcanic ash to the fine sediments. The rise of an ancestral Sierra

*Ghost Ranch in northern New Mexico is backed by Jurassic sandstone cliffs. Triassic mudstones below have yielded many specimens of* Coelophysis, *the earliest known dinosaur.*

Nevada far to the west cut off Pacific moisture, and the land grew dry—a Mesozoic Sahara.

Late in the era, the sea came once more, this time from the east as the center of the continent subsided. As the era closed, that sea, too, withdrew. And the dinosaurs died—just why we are not yet sure. Three quarters of the then-existing species of plants and animals became extinct at the end of Mesozoic time, possibly because of intense volcanic action or because of bombardment of the Earth by one or more large asteroids—either process raising so much dust that sunlight was dimmed or even virtually blotted out, perhaps for many years. As the Earth grew cold, plants died from lack of sun and rain. Animals died for want of food—the plant-eaters first, then the meat-eaters, and then the scavengers.

At the end of Mesozoic time, North America broke away from Europe and started on an epic voyage to the west. During the Cenozoic Era, the narrow rift between the continents widened into the Atlantic Basin as molten magma from the mantle welled up along the new mid-Atlantic Ridge and the sea floor spread apart. Westward movement of the continent has now lasted at least 66 million years—right up to the present time.

Voyaging westward, the North American plate collided with the East Pacific plate and overrode almost all of it, pushing it downward into the mantle. Small outlying islands and microcontinents of continental-type rock tacked themselves onto the leading edge of the North American plate. The continent buckled and broke, the modern Rocky Mountains rose in almost the

29

same position as their Paleozoic ancestors, and another round of volcanoes belched fiery rock and volcanic ash. As newborn streams fed newborn rivers, the mountains shed vast quantities of rock material, much of which was carried no farther than intermountain basins.

During this time new animals came to roam the Earth—the mammals. Here in New Mexico many varieties lived and died—some certainly the victims of showers of volcanic ash, some dying in floods from new-sprung mountains. Their fossilized remains, found in many parts of New Mexico, document the evolution of horses and camels, of wild dogs and large cats and bears, and of many species that left no modern descendants.

Two new geologic features began to develop about 30 million years ago—a pair of fault zones cutting north-south across New Mexico, from Colorado to Texas. Caused perhaps by the birth of a new line of mantle upwelling, the faults signaled the beginning of the Rio Grande Rift, a long sliver of crust that dropped down between the two lines of faults. At the same time, movement on other faults created steplike plateaus in northern Arizona, northwestern New Mexico, and adjacent Utah and Colorado.

Later in the era, about 15 to 8 million years ago, increasing tension in the crust stretched parts of the crust to the breaking point. In much of the Southwest, including southern New Mexico, pull-apart or extension faults established the tilted ranges and deep basins of the Basin and Range province. At about the same time, intense volcanism along the west side of the Rio Grande Rift, with great explosive eruptions that scattered volcanic ash over New Mexico and adjacent states, left gaping calderas where once stood giant composite volcanoes.

Throughout these changes, drainage patterns changed as well. The Rio Grande, born in the Colorado Rockies, at first flowed into closed basins along the great rift, as did many other streams draining smaller ranges near the rift. As basins filled with rock debris, rivers and streams joined up, until the Rio Grande became a through-flowing river carrying Rocky Mountain water all the way to the Gulf of Mexico.

In the fluctuating climates of the Ice Ages, when great glaciers covered much of North America and Europe, small mountain glaciers developed in the Brazos and Sangre de Cristo Mountains. The rest of New Mexico received much more precipitation than it does today. Lakes developed in many of the closed valleys of southern and central New Mexico. Floods were frequent, and vast amounts of rock material were eroded from the mountains and carried into intermountain basins, much of it building up as large alluvial fans along the valley margins. With rainy cycles that reflected ice advances far to the north, streams and rivers of New Mexico alternately deposited and cut down through thick layers of sand, gravel, and silt, leaving fragments of old floodplains as terraces along their borders.

A million years ago, with much of the landscape looking very much as it does today, another volcano—farther north than the others—exploded and collapsed into its magma chamber, creating the Jemez Caldera.

Since then, the Rio Grande Rift has continued to widen, mountains have shed more debris into valleys, rivers have picked up that debris and carried it eastward and southward. And a few smaller volcanic eruptions, some of them within the last few thousand years, have added piled-up cinder cones and flows of dark basalt, final accents witnessed by some of North America's early human inhabitants, Clovis and Folsom Man and their contemporaries and successors.

*Post-1880 erosion has deeply trenched soft floodplain deposits in many parts of New Mexico.* G.O. Bachman photo courtesy of U.S. Geological Survey.

31

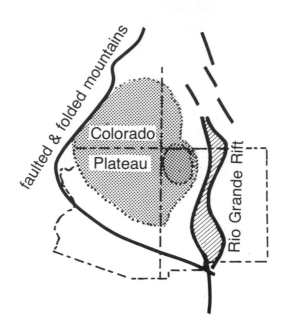

*The Colorado Plateau (light shading) is an almost undeformed block between a western fold belt and the Rio Grande Rift. Dark oval is the San Juan Basin.*

# II
# The Western Highlands

The region west of the Rio Grande Rift divides scenically as well as geologically into plateau country to the north and rugged but beautiful volcanic mountains to the south. The plateau country is the southeastern corner of the great Colorado Plateau of northern Arizona, western Colorado, and southern Utah, named for the river which, with its tributaries, cuts magnificent canyons through flat-lying sedimentary rocks. In New Mexico these rocks range in age from Pennsylvanian to Tertiary.

Most of the plateau country consists of large blocks of horizontal or nearly horizontal sedimentary rock underlain by Precambrian granite and gneiss. However the Precambrian rocks are exposed only in the depths of Grand Canyon and along the far western edge of the whole plateau region. Bordered by faults or by simple monoclines that in many—perhaps all—cases drape over faults in Precambrian rock, the plateaus vary in height and width. Along the Arizona-New Mexico line, the rocks of eastern Arizona fold down sharply in a prominent monocline—a feature that can best be seen along Interstate 40 near Gallup—so that in the plateau country of New Mexico the oldest sedimentary rocks are not exposed at the surface.

Northeast of Gallup the downward bend extends under the San Juan Basin, an almost circular basin 110 miles wide, somewhat out of keeping with the uplifted blocks that characterize the rest of the plateau country. Outlined with Cretaceous sedimentary rocks, the basin is centered with soft Tertiary sandstones and siltstones in more or less horizontal position. This is a barren region little known and little understood until oil and gas were discovered below its surface. The basin is now spangled with wells, some with slowly rocking pumps but many marked only by "Christmas trees" of pipes and valves, and by clustered oil and gas storage tanks.

Along the eastern edge of the San Juan Basin, underlying sedimentary rock layers rise sharply to the surface, the harder layers standing out as hogback ridges bordering the Brazos and Nacimiento uplifts—high, much faulted blocks of Precambrian and Paleozoic rocks.

The western and southeastern parts of the basin are dotted with gaunt volcanic necks—the eroded hearts of old volcanoes. The most famous is Shiprock, west of Farmington, rising 1100 feet above its surroundings, with radiating dikes forming walls and ridges across the desert: a "textbook" volcanic neck. A swarm of dikes marks the northeastern corner of the basin near Lumberton.

The Chuska Mountains west of the San Juan Basin are a long ridge-like anticline topped with Triassic, Jurassic, and Tertiary rock. Absence of Cretaceous rocks, which elsewhere come between Jurassic and Tertiary layers, shows that the range first lifted its head in Mesozoic or early Tertiary time. The mountains are fringed with landslides where younger rocks slid down over soft, slippery shales that wall the range.

Farther south, but along the same northwest-southeast trend as the Chuskas, are the Zuni Mountains. They are also an anticline—one that pushed upward to even greater heights

| | basin, down-faulted block | | caldera | | major fault |
|---|---|---|---|---|---|

*New Mexico's Western Highlands consist of part of the Colorado Plateau (whose dominant feature here is the San Juan Basin), the Datil-Mogollon volcanic highlands, and a part of the Basin and Range region in the south.*

than the Chuskas. Erosion has stripped away all the sedimentary rocks that used to cover the Zuni Mountains, exposing their Precambrian granite core. In the Zunis, Pennsylvanian sedimentary rocks were deposited right on this ancient granite. Older sedimentary rocks—Cambrian to Mississippian—once existed here but were stripped away during Pennsylvanian time, when this region rose as part of the ancestral Rockies.

A broad field of fairly recent lava flows and cinder cones, as well as a few outlying mesas, lies south of the Zuni Mountains and forms the border of the plateau country. Southward the mesas abut the volcanic jumble of the Datil-Mogollon Highlands, where Tertiary volcanoes poured out vast quantities of silicic lava and volcanic ash. Some of the great volcanoes exploded during eruption and collapsed into their magma chambers, leaving giant calderas as much as 30 miles across. Though they are now deeply eroded, careful geologic study has outlined many of the old calderas.

Several fault-edged sunken blocks—San Agustin Plain among them—cut across the Datil-Mogollon area. Near the southern end of the region, slender fault slices of Paleozoic and Cretaceous rocks remain to tell us a little about the earlier history of the area. Paleozoic rocks are in places highly mineralized, making this one of the chief mining areas of the state. On the east, the highlands give way to the fault-edged Rio Grande Rift.

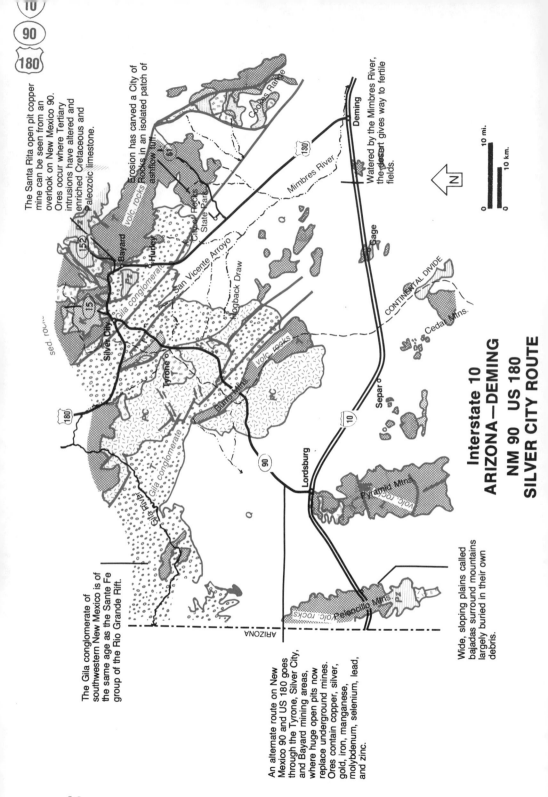

The Santa Rita open pit copper mine can be seen from an overlook on New Mexico 90. Ores occur where Tertiary intrusions have altered and enriched Cretaceous and Paleozoic limestone.

Erosion has carved a City of Rocks in an isolated patch of ashflow tuff.

Watered by the Mimbres River, the desert gives way to fertile fields.

The Gila conglomerate of southwestern New Mexico is of the same age as the Sante Fe group of the Rio Grande Rift.

An alternate route on New Mexico 90 and US 180 goes through the Tyrone, Silver City, and Bayard mining areas, where huge open pits now replace underground mines. Ores contain copper, silver, gold, iron, manganese, molybdenum, selenium, lead, and zinc.

Wide, sloping plains called bajadas surround mountains largely buried in their own debris.

**Interstate 10**
**ARIZONA—DEMING**
**NM 90   US 180**
**SILVER CITY ROUTE**

# I-10
# Arizona—Deming
## 86 mi./138 km.

Most of the rocks seen in the first few miles east of the Arizona border, in the Peloncillo Mountains, are volcanic—primarily Tertiary tuff and lava flows. Near the ghost town of Steins near milepost 3, these rocks were for a time quarried and crushed for use as ballast for the Southern Pacific Railway. Foundations of the old mill still stand.

Leaving the mountains, the highway closely follows the route of the Butterfield Trail, in pre-railroad days a stagecoach route to California. It descends the broad bajada or alluvial apron that surrounds the mountains. Bajadas are common in the Southwest; they form as alluvial fans grow and merge with their neighbors, and as mountain faces "retreat" from the faults along which uplift occurred.

Near milepost 5, the route crosses three beach ridges, unfortunately obscured near the highway, of Pleistocene Lake Animas—a lake that was once 17 miles long and about 50 feet deep. Lake Animas reached its greatest extent 12,000 years ago during the last Ice Age, and may still have been here as late as 4000 years ago. At its greatest extent, the lake was probably fresh; it is thought to have drained northward toward the Gila River Valley. As its waters evaporated, bringing lake level below outlet level, the lake became increasingly salty. A broad modern playa, with deposits of salt and other evaporite minerals, now occupies part of the old lake floor. Like many other valleys in the Basin and Range region, this is still a closed basin, and does not drain. Water entering it flows toward its lowest spot and there remains, slowly evaporating, leaving its dissolved minerals— salt, gypsum, sodium carbonate, and other compounds—behind on the playa surface.

When dry, the playa is marked with a network of mudcracks, the result of the shrinking of the clayey sediments each time they dry. Over much of the lake surface the mudcrack pattern is quite small-scale, but out near the center of the playa huge cracks as much as a foot wide mark out 10-foot or larger polygons arranged in long parallel rows.

In the southern part of this valley, south of Interstate 10 near the town of Cotton City, there is a geothermal area where naturally heated water lies close to the surface, warming the ground enough to cause rapid snow melt (when and if it snows) and earlier than normal springtime greening of desert and crops. Some uranium has been found in nearby ranges.

Continuing eastward, the highway crosses part of the playa and, near mileposts 12 and 13, once more cuts through Pleistocene beach ridges—two of them exposed in a low roadcut marked by yuccas, which seem to prefer the sandy soil of the beach ridges. The highway then climbs gradually across the bajada surrounding the Pyramid Mountains. Except for intrusive rocks near the north end of the range, these mountains are composed almost entirely of Tertiary volcanic rocks, some of it volcanic tuff associated with huge volcanic explosions and collapse of calderas to the south and west.

In 1870, silver was discovered in the Tertiary intrusion near the north end of the Pyramid Range, triggering a local mining boom. The initial boom was a hoax—ores were not nearly as rich as claimed. Two years later another and more deliberate hoax was fomented when promoters "salted" a small area with Brazilian and African diamonds, hoping to impress investors. There *is* silver here—it was mined at Shakespeare, now a ghost town, until 1885—but there are no diamonds. Copper, gold, lead, zinc, and uranium have also been mined in the Pyramid Mountains, which get their name from the dark conical peak, a volcanic neck, at the north end of the range.

East of Lordsburg the country becomes less mountainous. Immediately east of town the interstate crosses Lordsburg Draw, once an arm of Lake Animas, and a place of geologic interest because of discovery and excavation of a mammoth skeleton nearby.

Northeast of Lordsburg are the Burro Mountains, fault slices of Precambrian granite, Tertiary volcanic rocks, and Gila conglomerate—a Tertiary rock derived, as stream gravel, from the many different ranges adjoining the upper drainage basin of the Gila River. There are copper-lead-zinc mines in the Precambrian part of the Burro Mountains, some of them large open pit mines.

Small ranges and clustered hills south of the highway are fault blocks of Tertiary volcanic rocks and Paleozoic sedimentary

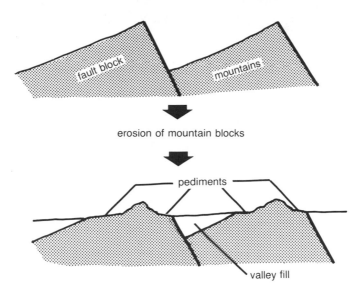

*Pediments form at mountain bases as erosion pares away the ranges and fills the valleys. Later erosion of the valley fill may leave the pediments as high shelves.*

rocks—more Basin and Range fault block mountains, with only their tips exposed above the thick valley fill.

The town of Playas, about 20 miles south of the highway, is a smelter town where ores mined at Tyrone, near Silver City, are processed. When built, the smelters at Playas were among the most modern in the world; they process some 2000 tons of concentrated ore per day.

Interstate 10 crosses the Continental Divide at an elevation of 4,500 feet. The divide—the continent's "backbone"—separates streams that flow west to the Pacific (in this case via the Gila and the Colorado rivers to the Gulf of California) from those that flow east to the Atlantic (via the Rio Grande to the Gulf of Mexico). The divide is scarcely discernable here, and there are no streams to separate. East of it, the highway descends an imperceptible slope among scattered hills and small ranges that are again only the summits of deeply buried ranges.

Several of the little ranges contain Paleozoic limestones intruded by Tertiary granite. Along the contact, mineral enrichment was catalyzed by the calcium carbonate of the limestone. Silver, lead, zinc, and gold were produced from these ranges, mostly in the late 19th century. Near milepost 65 are the excavations of a quarry from which limestone was obtained for construction of this part of Interstate 10.

*Ashflow tuff from Tertiary volcanoes has weathered and eroded into massive "buildings" at City of Rocks State Park. These are outposts of the main cluster of rocks.*

Recently, several oil exploration wells were drilled a few miles north of the highway here. No commercial amounts of oil or gas were discovered, but the wells did produce some interesting geologic information: Precambrian rocks lie 7000 to 8000 feet below the surface, and there is a major unconformity between lower Paleozoic and upper Cretaceous sedimentary rocks, with all of later Paleozoic time and most of Mesozoic time missing from the record. In the Burro Mountains to the northwest, there are no Paleozoic rocks at all; Cretaceous rocks rest right on Precambrian granite. We are sure the missing sediments were deposited here, however, so we deduce that in Mesozoic time this area must have been lifted high enough for erosion to remove the missing rock layers. Toward the south, the rock record is more complete: Paleozoic and Cretaceous sedimentary rocks are more than 8000 feet thick, and the Precambrian surface is deep underground.

Farms of the Mimbres Valley near Deming are nourished by deep wells tapping groundwater that flows from the mountains to the north. The valley was settled in the 1880s by farmers brought here by mining companies to provide groceries for their miners. At that time all irrigation was from the Mimbres River. The Mimbres Basin has no external drainage—the Mimbres and other streams simply sink into the valley fill. Water levels are now dropping because pumping for irrigation exceeds natural recharge. Pumping of groundwater has caused local subsidence of the desert floor and cracking or fissuring of its surface.

The Little Florida Mountains southeast of Deming are the site of Rock Hound State Park, unusual among state parks in that rock collecting is not only allowed but encouraged! Crystal-centered geodes and colorful agates, formed by groundwater depositing minerals in rock cavities, occur among volcanic rocks there.

*Capitol Dome at the north end of the Florida Mountains is capped with Tertiary volcanic rocks lying over Paleozoic sedimentary rocks. Low knob at left is Precambrian granite.* N.H. Darton photo courtesy of U.S. Geological Survey.

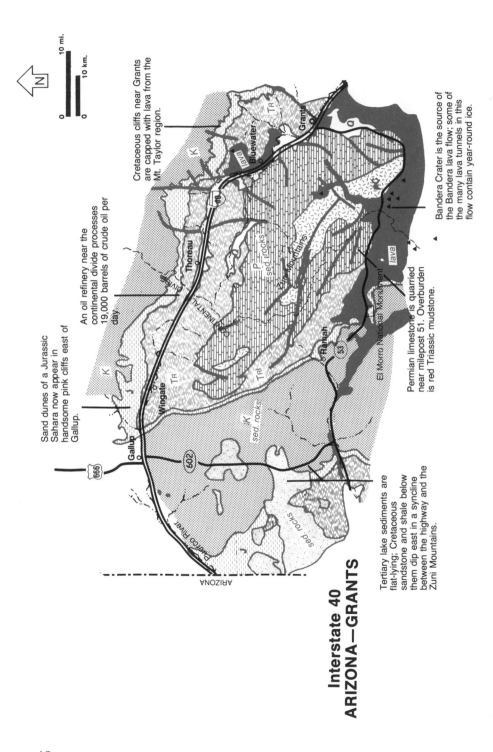

**Interstate 40**
**ARIZONA—GRANTS**

Cretaceous cliffs near Grants are capped with lava from the Mt. Taylor region.

An oil refinery near the continental divide processes 19,000 barrels of crude oil per day.

Sand dunes of a Jurassic Sahara now appear in handsome pink cliffs east of Gallup.

Bandera Crater is the source of the Bandera lava flow; some of the many lava tunnels in this flow contain year-round ice.

Permian limestone is quarried near milepost 51. Overburden is red Triassic mudstone.

Tertiary lake sediments are flat-lying; Cretaceous sandstone and shale below them dip east in a syncline between the highway and the Zuni Mountains.

N

0 ——— 10 mi.

0 ——— 10 km.

*Tall buttresses of resistant Jurassic sandstone add their brilliant color to the route east of Gallup. The foreground valley, eroded in weak Triassic and Jurassic shale and mudstone, is floored with fine silt now etched with gullies.*

# I-40
# Arizona—Grants
### 79 mi./127 km.

Following the valley of the Puerco River, this highway enters New Mexico between cliffs of tan Jurassic sandstone. Large-scale crossbedding, along with fine, well-sorted sand grains, points to the formation's origin as sand dunes of a long-ago desert. The sandstone weathers to smooth surfaces with many cavities and arching alcoves where unsupported rock has fallen away.

Above the main cliffs are younger Mesozoic rocks, notably the Dakota sandstone, deposited on a Cretaceous beach, now topping the cliffs. Except for several small anticlines, these rock layers all dip eastward, so younger and younger rocks appear as we proceed in that direction. Within a few miles the massive Jurassic sandstone disappears and reddish Gallup sandstone, also Cretaceous, tops the cliffs.

The Puerco River flows (when it flows) in a deep gully etched in fine, tan, poorly consolidated silt of its own floodplain. Sediments such as these commonly contain charcoal, charred animal bones and teeth, and other mementos left by prehistoric inhabitants. The carbon of the charcoal can be dated accurately by the carbon-14 method, which works through the slow, steady transformation of radioactive carbon-14 to nitrogen.

Roadcuts near Gallup expose crossbedded sandstone and thin beds of coal, with some ancient sand-filled channels cutting into the coal. The coal formed from plant material deposited in marshy or swampy lagoons along Cretaceous shores. Layers of both sandstone and coal thin and thicken and disappear—the irregularities one would expect in the coastal environment in which these rocks were deposited. Coal deposits near Gallup were among the factors considered in routing the Santa Fe Railroad here.

East of Gallup along Interstate 40, old cars have been dumped into a deep gully of the Puerco River in an effort to slow or stop erosion of the thick, soft silt of the former floodplain.

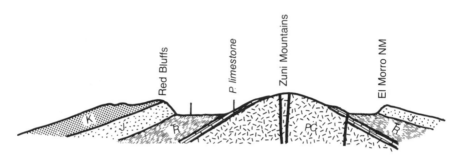

*Section across I-40 near Thoreau*

Cretaceous and older rocks bend up suddenly just east of Gallup, so that from this point eastward, successively *older* rocks—Cretaceous, then Jurassic, then Triassic—border the highway. This is the west side of the Zuni Uplift, an oval-shaped dome whose oldest rocks, 10 to 15 miles south of the highway, form the Zuni Mountains. With huge cliffs of Jurassic Entrada sandstone to the north, the highway runs to Grants and beyond in a racetrack valley eroded by the Puerco River in soft Triassic mudstone and siltstone. Some of the Triassic rocks below the massive Jurassic sandstone appear in highway roadcuts. As these soft rocks erode, undermining the Entrada sandstone, great slabs of the cliffs fall away and the valley gradually widens. The cliffs slowly retreat northward.

Shortly before crossing the Continental Divide near milepost 48, the highway passes an oil refinery. At the divide, we leave the Pacific drainage of the Colorado River basin, and enter the Atlantic drainage of the Rio Grande. Ahead to the northeast is the graceful volcanic cone of Mt. Taylor, discussed in the next section.

The red cliffs of Entrada sandstone end abruptly near Prewitt, where Jurassic rocks are cut off by a large fault. To the east are

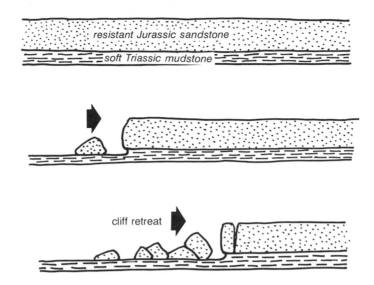

*As erosion of soft siltstone undermines sandstone cliffs, large blocks fall away. Over time, the cliffs retreat.*

lava-capped cliffs of tan sandstone and gray shale of the Mesaverde group, originally deposited in a Cretaceous sea. They are warped into shallow anticlines and synclines. In places, faults confuse the geologic picture. Just southeast of Prewitt the highway crosses a Quaternary lava flow, the youngest rock you will have seen along this route.

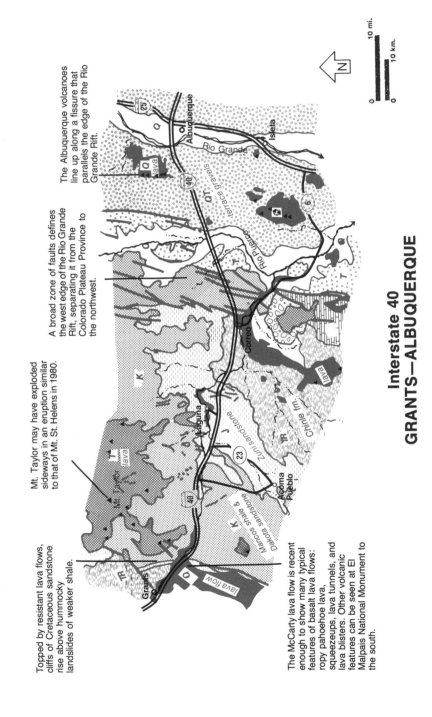

Topped by resistant lava flows, cliffs of Cretaceous sandstone rise above hummocky landslides of weaker shale.

Mt. Taylor may have exploded sideways in an eruption similar to that of Mt. St. Helens in 1980.

A broad zone of faults defines the west edge of the Rio Grande Rift, separating it from the Colorado Plateau Province to the northwest.

The Albuquerque volcanoes line up along a fissure that parallels the edge of the Rio Grande Rift.

The McCarty lava flow is recent enough to show many typical features of basalt lava flows: ropy pahoehoe lava, squeezeups, lava tunnels, and lava blisters. Other volcanic features can be seen at El Malpais National Monument to the south.

**Interstate 40**
## GRANTS–ALBUQUERQUE

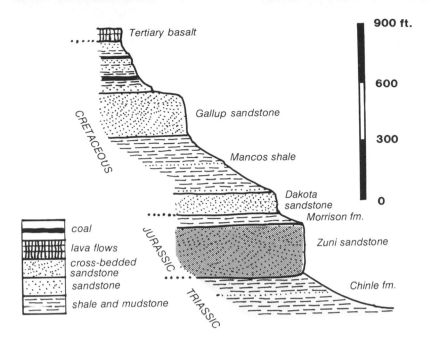

*This stratigraphic diagram shows strata near I-40 between Arizona and Grants. The lower part of the section is visible east of Gallup, the upper part east of Grants.*

# I-40
# Grants—Albuquerque
### 72 mi./116 km.

Grants lies in a flat-floored valley walled on the south by lava flows and on the north by cliffs of Jurassic and Cretaceous sedimentary strata capped with lava flows. A low, almost imperceptible anticline runs north-south through the town and is reflected in the gentle arch of these sedimentary layers. Steep, hummocky slopes along the base of the cliffs are landslides in which lava and harder Cretaceous rock have slid down over soft, slippery Cretaceous shale.

For many years Grants has been a major center for uranium exploration and mining, with mines mostly in the Morrison formation, the youngest Jurassic formation in this area. Uranium accumulates as water, filtering through volcanic or intrusive rocks, picks up soluble uranium minerals and redeposits them wherever it finds some organic matter—plant- and animal-derived materials. The Morrison formation, deposited on a river floodplain in richly vegetated surroundings, consists of variously colored mudstone containing lots of clayey bentonite derived from volcanic ash. It also contains some thin layers of orange sandstone, stream-channel deposits rich in organic matter, within which much of the uranium concentrated. The source of the uranium is believed to be either intrusive rocks in the Zuni Mountains to the south, or volcanic ash of the Morrison and other formations.

The Grants uranium district extends eastward to Laguna. Many of the mines worked during the uranium booms of the 1950s and '70s are closed now. Some were mined out, but others are "on hold," waiting for a day when the price of uranium rises.

*The Grants mineral belt extends nearly 100 miles along the southwest flank of the San Juan Basin. Uranium occurs in a band 10 to 20 miles wide, in filled-in channels of Jurassic streams.*

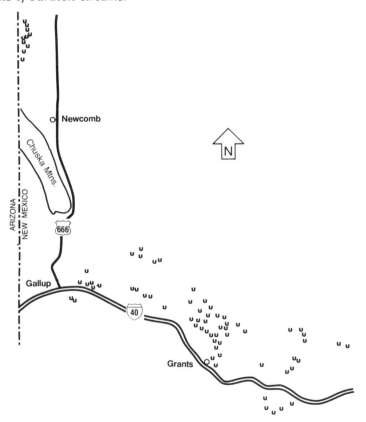

A large lava field known as the Malpais reaches the highway east of Grants. The lava flows came from small volcanoes 20-30 miles southwest of the highway, in El Malpais National Monument. The youngest flow is less that 1000 years old, and may figure in Indian

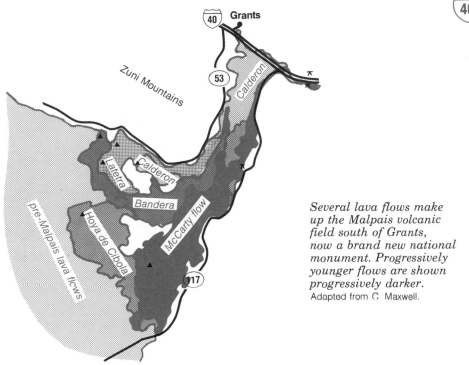

*Several lava flows make up the Malpais volcanic field south of Grants, now a brand new national monument. Progressively younger flows are shown progressively darker.*
Adapted from C. Maxwell.

legends as "fire rock" that buried the fields of Indian ancestors. Stop at the rest areas near milepost 93 for a look at the tip of this flow. Its jumbled, twisted, ropy lava is full of vesicles, gas-bubble holes, and the flow as a whole is marked with pressure ridges, squeeze-ups, and grooved lava surfaces that show where molten lava pushed through its hardening crust. Water-filled sags represent collapsed lava tubes that formed where lava flowed out from under its own hardening crust. The geologist who in 1946 first described this fresh-looking

*A domelike lava blister has cracked open, forming a sharp ridge across part of the McCarty Lava Flow.*

49

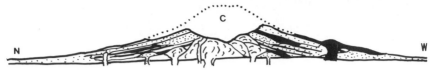

*Mt. Taylor, with its alternating lava flows (black) and volcanic debris (dotted), may have had a history similar to that of Mt. St. Helens in Washington. A central lava dome, partly destroyed by erosion, is the youngest part of the mountain.* Adapted from Crumpler, 1982.

flow also coined many of the terms now used to describe lava flow features.

The highway proceeds eastward down the valley of the Rio San Jose. Basalt-capped mesas on either side expose on their flanks Cretaceous sandstone still dipping east off the Zuni uplift. Note the columnar joints of the basalt.

From the rest area at milepost 102, Mt. Taylor (11,301 feet) shows up to the north, towering over a lava-capped mesa. Mt. Taylor is built of many successive lava and ash flows, the oldest of basalt, the younger ones of dacite and andesite. The mountain began to form about four million years ago, and for two million years erupted repeatedly, building stiff lava domes, disgorging lava flows, and shooting out huge hot clouds of volcanic ash. Steaming mudflows at times poured down the mountainside in scenes that bring to mind the Mt. St. Helens eruption of 1980 in Washington state. The volcano today is somewhat horseshoe shaped, with an interior valley on its eastern side, suggesting further analogy with Mt. St. Helens and its sideways explosion. The volume of volcanic debris around the mountain suggests that more than once the volcano destroyed itself, only to rebuild. However, the last eruption occurred more than two million years ago, so it is unlikely that the volcano will awaken again.

The highway continues eastward through Cretaceous rocks along the Rio San Jose. The boggy area between the highway and the pueblo of Laguna, near milepost 114, is the original "lake" that led to

*Laguna Peublo rises against lava-capped mesas crowned with distant Mt. Taylor, an eroded composite volcano.*

*Only partly protected by hard caprock, Zuni sandstone of northwestern New Mexico erodes into sculptured pinnacles.* W.T. Lee photo courtesy of U.S. Geological Survey.

the Spanish name for the town. East of this, the route crosses another lava flow, older than those of the Malpais, and covered with soil and wind-blown sand.

To the north, lower mesas appear below the Cretaceous cliffs. They are capped with Jurassic rocks—the Entrada sandstone—and their slopes are formed in the soft layers of the Morrison formation, also Jurassic. Still farther east, soft red-tinted Triassic mudstone forms the valley floor.

Between mileposts 130 and 140 we leave the flat-topped tablelands of the Colorado Plateau and enter a broad fault zone that defines the western edge of the Rio Grande Rift. Although the faults here cause disruptions in Mesozoic and Tertiary sedimentary layers, they are hard to see from the highway. But because of the faults, east of the Rio Puerco there are no more Mesozoic rocks exposed on the surface. Soft, poorly consolidated sediments between here and Albuquerque are either partly compacted gravels and sand of the Santa Fe group, carried here by the Rio Grande before it deepened its present valley, or recent gravel and sand washed from nearby mountains.

Broad, gently sloping terraces edge both the Rio Puerco and the Rio Grande valleys, each terrace representing a time of balance between the depositing of the Santa Fe group and downward erosion. The uppermost terraces are the oldest, with renewed erosion cutting new channels, which in turn fill in with river floodplain deposits that will

51

end up as lower terraces. The Santa Fe group varies in thickness from 1000 to 6000 feet and includes many layers of volcanic ash and lava.

Between mileposts 145 and 150, the highway crosses a particularly broad terrace whose irregular, bumpy surface was a sand dune field in Pleistocene time. The small Albuquerque volcanoes can be glimpsed to the north—a string of five small cones lined up from north to south along a fault or fissure. The oldest lava flows surrounding these volcanoes have been dated at about 190,000 years.

From the rest area east of milepost 151, the city of Albuquerque, resting on several terrace levels, is laid out before you. Most of the city is east of the Rio Grande, below the steep-cliffed Sandia Mountains, which form the other side of the Rio Grande Rift. The rift valley is about 28 miles wide here.

The Rio Grande has been channelized where it passes through Albuquerque, with floodways on either side of the main channel. The inner valley now occupied by the river is a newcomer in terms of its geologic age. Cutting of the inner valley began several hundred thousand years ago, and was followed perhaps as recently as 20,000 years ago by the present cycle of renewed deposition, which has created about 75 feet of new valley fill.

As it swung from side to side, silting up its channel and shifting its course, the river formed natural levees that gradually raised the river level higher than its floodplain. Lowlands on either side of the river became swampy and subject to flooding, particularly when tributaries added runoff from sudden severe rainstorms. To prevent floods, diversion channels have been built along the mountain front to catch runoff and route it downstream. The danger of flooding of valuable bottomland is further lessened by flood-control dams on several upstream tributaries.

*Dry gulches shaped by cloudburst-swollen streams are common in rugged country west of Socorro. The walls of this gulch are composed of Tertiary breccia, broken volcanic material from the explosion that created the Socorro caldera.*

# US 60
# Socorro—Datil
### 63 mi./101 km.

Socorro lies at the northeast corner of the Datil-Mogollon volcanic field, and has as a backdrop several volcanic peaks. The city developed as a mining and smelter town in the 1880s, soon after the discovery of silver and silver-lead ores on Socorro Peak and near Magdalena, about 30 miles to the west. By 1889 the main smelter in Socorro was producing 300 bars of silver per day, extracted from 250 tons of ore, some of the ore shipped in from Arizona, Colorado, and Mexico. In 1882 the town became the home of the New Mexico School of Mines, now the New Mexico Institute of Mining and Technology. The New Mexico Bureau of Mines and Mineral Resources is headquartered on campus.

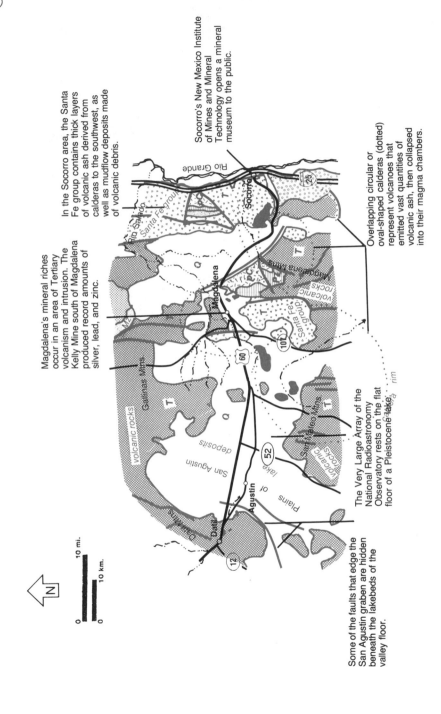

Magdalena's mineral riches occur in an area of Tertiary volcanism and intrusion. The Kelly Mine south of Magdalena produced record amounts of silver, lead, and zinc.

In the Socorro area, the Santa Fe group contains thick layers of volcanic ash derived from calderas to the southwest, as well as mudflow deposits made of volcanic debris.

Socorro's New Mexico Institute of Mines and Mineral Technology opens a mineral museum to the public.

Overlapping circular or oval-shaped calderas (dotted) represent volcanoes that emitted vast quantities of volcanic ash, then collapsed into their magma chambers.

The Very Large Array of the National Radioastronomy Observatory rests on the flat floor of a Pleistocene lake.

Some of the faults that edge the San Agustin graben are hidden beneath the lakebeds of the valley floor.

N

0    10 mi.

0    10 km.

## US 60
## SOCORRO—DATIL

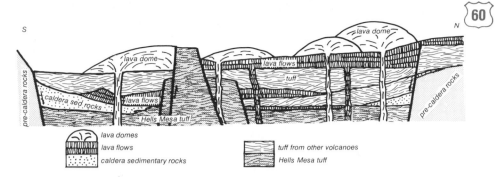

S    N

pre-caldera rocks

lava dome

lava flows

tuff

Hells Mesa tuff

caldera sed rocks

lava flows

pre-caldera rocks

lava dome

lava flows

*lava domes*

*lava flows*

*caldera sedimentary rocks*

*tuff from other volcanoes*

*Hells Mesa tuff*

*Eruption of the Hells Mesa Tuff brought about the collapse that created the Socorro caldera. Note the uplifted central blocks. The present profile of the Socorro and Chupadera Mountains, where the caldera occurs, is not shown in this reconstruction. After Chamberlain and Eggleston.*

Mountains west and southwest of town form a rugged region of eroded Tertiary volcanoes and their products. Volcanism in this region dates back to early Tertiary time, but most of the volcanic rock visible in the mountains—lava flows, lava domes, and voluminous amounts of tuff—formed in Oligocene to Miocene time. Explosive eruptions far greater than any in historic time caused the collapse of a number of large volcanoes, creating many basin-like calderas. These calderas can still be recognized by mapping the ring-shaped faults, sudden thickening of sheets of tuff, alluvial fans, mudflow remains, and lake deposits associated with them.

Socorro Peak, the one with the M, is at the margin of one of the calderas. On its slopes, tilted, light-colored Pennsylvanian limestone and shale are overlain by volcanic breccia and volcanic ash deposited in the caldera. These deposits are overlain by Miocene volcanic rock that forms the summit of the peak. Other caldera rocks are visible as the highway continues into the mountains.

Rounding the southern end of the Socorro Mountains, the highway passes a pumice mine. Foamy, light-colored volcanic glass, pumice is

*A dike forms a steep-sided ridge bordered with fertile volcanic soils.*

used for lightweight aggregate, as an abrasive and polish, and as a soil-lightener for greenhouses and houseplants.

Silver was discovered near Magdalena in the 1880s, but the ore was of poor quality and was not mined at the time. However, Kelly, about three miles south of Magdalena, was the site of a real lead-silver boom. Ore bodies in this region occur in Paleozoic limestone altered and enriched by Tertiary intrusions—a common mode of occurrence as limestone provides the special chemical environment needed to help the formation of ore minerals from metal-bearing fluids that accompany intrusions. The Magdalena District, including the Kelly mines, shipped $7,000,000 to $9,000,000 worth of lead-silver ore between 1880 and 1902. Iron ores were later found a few miles west and north of Magdalena, in dark brown nodules just lying around on the surface! The nodule supply was soon exhausted, however, and mines were opened nearby.

Both Magdalena and Kelly offer good mineral collecting. Azurite, barite, malachite, cerussite, pyrite, and other minerals can be found in the old mine dumps. Most of the dumps are private property, however, and permission should be obtained to search them.

The broad valley west of the Magdalena Mountains is San Agustin Valley, or the Plains of San Agustin—a beautiful flat-floored, mountain-bordered valley once the site of a 50-mile-long Pleistocene lake. Geologically a sunken block that subsided between parallel faults, the valley is now the site of a radioastronomy observatory, where a "Very Large Array" of saucerlike antennae gathers radio signals from far out in interstellar space. Signals emitted millions and even billions of years ago, only now reaching the Earth, give scientists here a look at conditions within the universe at the time

*Saucer-shaped receivers of the National Radio Astronomy Observatory tower above the Plains of San Agustin. The observatory seeks answers to the origin of the universe.*

when our solar system was young. A visitor center, slide show, and walking tour explain the various areas of study—among them the births and deaths of stars, the properties of galaxies like our own, and the history of the origin of the universe. Most of the array is on the old lake floor, which facilitates contracting and expanding the array by moving the towers along intersecting railroad tracks.

San Agustin Valley was once on the receiving end of a meteor shower: Many meteorite fragments have been found on its surface.

The town of Datil lies at the northwest edge of the Plains of San Agustin, below more volcanic mountains. Datil was formerly a mining and smelter town.

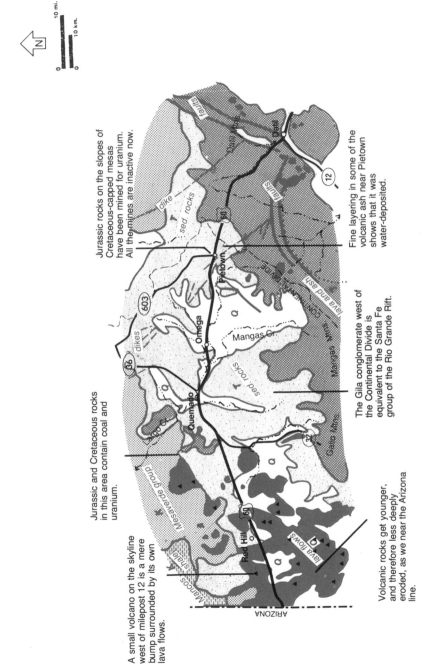

Jurassic rocks on the slopes of Cretaceous-capped mesas have been mined for uranium. All the mines are inactive now.

Fine layering in some of the volcanic ash near Pietown shows that it was water-deposited.

Jurassic and Cretaceous rocks in this area contain coal and uranium.

The Gila conglomerate west of the Continental Divide is equivalent to the Santa Fe group of the Rio Grande Rift.

A small volcano on the skyline west of milepost 12 is a mere bump surrounded by its own lava flows.

Volcanic rocks get younger, and therefore less deeply eroded, as we near the Arizona line.

## US 60
## DATIL—ARIZONA

*Ashflow tuff sometimes erodes into pinnacles and residual boulders that from a distance resemble outcrops of granite. Closer inspection is needed to identify these very different rock types correctly.*

# US 60
# Datil—Arizona
## 78 mi./125 km.

As the highway heads into the mountains west of Datil, we see dramatic boulders of pinkish gray and lavender volcanic ash. Much of this rock is a tightly welded ashflow tuff, widespread here near its source in Tertiary volcanoes of the Datil-Mogollon volcanic field. Cut by more or less vertical shrinkage joints that formed as it cooled, the rock in places erodes into free-standing pinnacles. Some unwelded ashfall tuff occurs here as well. You can be sure that where highway cuts are nearly vertical you are looking at welded or ashflow tuff; where they are slanted way back you are dealing with softer, slide-prone ashfall tuff.

Farther west, near Pietown, similar rock is stratified, and looks like sedimentary rock. It is composed of volcanic ash that fell in a lake or stream, and since it is water-deposited, it should be considered a sedimentary rock, even though its constituents are entirely volcanic. The Mt. St. Helens eruption in 1980 showed just how important water is in transporting and depositing volcanic ash; huge mudflows there carried enough material to block shipping channels in the Columbia

River. A lot of volcanic ash is foamy and will float for long distances on or suspended in the raging waters that originate in melted snows or in the heavy clouds that form above erupting volcanoes.

At milepost 58, the highway crosses the Continental Divide, which separates stream drainages that ultimately lead to the Atlantic (via the Rio Grande and the Gulf of Mexico) from those that lead to the Pacific (via the Colorado River and the Gulf of California). The divide runs southeastward along the crest of the Mangas Mountains. Going north it swings sharply westward across a broad saddle between the Mangas and Zuni mountains. West of the divide are more volcanic sedimentary rocks, as well as more welded ashflow tuff.

The peak to the southwest between Pietown and Omega is Escondido Peak, a Tertiary volcano reshaped by erosion. A remarkable sedimentary plain extends westward from it, formed by coalesced alluvial fans around its base. In Tertiary time the plain was continuous with similar surfaces flanking adjacent ranges and extending into Arizona. The Gila conglomerate of which it is made is equivalent in age to the Santa Fe group of the Rio Grande Valley. Only remnants of the old surface remain—relics of an earlier time and a lower elevation. Uplift of the entire Southwest region caused the streams that deposited these extensive gravels to cut down through them, carrying their pebbles, cobbles, and sand westward to fill in much lower areas.

Some of these Tertiary sediments, tilted by late Tertiary folding and faulting, are exposed near the highway between Omega and Quemado.

West of Quemado the highway climbs up onto the Tertiary surface, remaining on it until we encounter younger volcanic rocks at the eastern edge of Arizona's White Mountain volcanic field. Quaternary volcanoes in this region erupted basalt lava, darker and more fluid than the Tertiary lavas that built up the Datil-Mogollon volcanic field in which we have been traveling. In relatively non-explosive eruptions, basalt lavas built up to sizable elevations in the White Mountains of Arizona; they also filled stream valleys and spread out in thin sheets that now cap many mesas. They are young enough not to be broken by faults or tilted and folded by earth movements. Basalt lavas originate in the Earth's mantle, below the crust, so we can be sure that where they are present there has been some very deep faulting.

To the north are mesas capped with Cretaceous sandstone belonging to the Mesaverde group. These essentially horizontal rocks mark the edge of the Colorado Plateau of northern Arizona, southern Utah, and adjacent parts of Colorado and New Mexico. The mesas capped with horizontal lava blend with those formed entirely of horizontal layers of sedimentary rock. The small volcanic centers from which the lava came appear as small cones on the tops of the lava caps.

As the highway climbs onto a basalt flow west of milepost 12 we can glimpse the red soil at its base, baked to brick by the heat of the molten rock.

In the last few miles before the Arizona line the highway crosses some more dark gray Quaternary volcanic rocks, quite different in character from the purple and reddish gray Tertiary volcanic rocks around Datil and Socorro. Bright brick red sedimentary rocks below some of the flows are Triassic, part of the Moenkopi formation of the Painted Desert to the northwest.

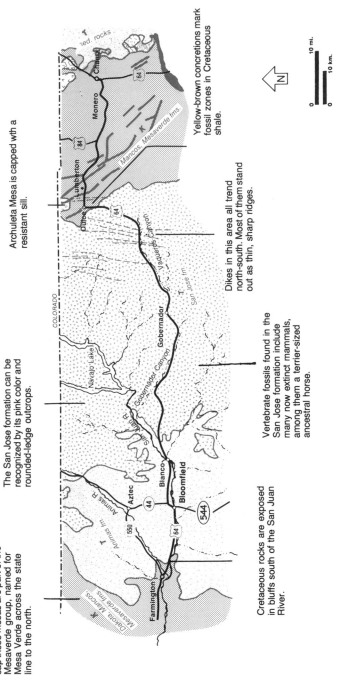

Yellow-brown concretions mark fossil zones in Cretaceous shale.

Archuleta Mesa is capped with a resistant sill.

Dikes in this area all trend north-south. Most of them stand out as thin, sharp ridges.

The San Jose formation can be recognized by its pink color and rounded-ledge outcrops.

Vertebrate fossils found in the San Jose formation include many now extinct mammals, among them a terrier-sized ancestral horse.

Cretaceous sandstones that cap these mesas are part of the Mesaverde group, named for Mesa Verde across the state line to the north.

Cretaceous rocks are exposed in bluffs south of the San Juan River.

## US 64
## CHAMA—FARMINGTON

# US 64
# Chama—Farmington
### 106 mi./171 km.

A short distance south of Chama, US 64 turns westward, skirting tall bluffs of Mancos shale capped with sandstone of the Mesaverde group. Both units are Cretaceous. The soft, dark gray Mancos shale—a marine shale—floors the Chama syncline, which extends south for about 30 miles. Largely a near-shore sandstone, the Mesaverde group also includes shales and seams of coal deposited as plant material in marshes and embayments of the Cretaceous coast.

Near milepost 348 we cross the Continental Divide, leaving the Atlantic or Rio Grande drainage and entering the Pacific drainage of the San Juan and Colorado rivers. Elevation of the divide here is 7776 feet. A small patch of Dakota sandstone, blocky and resistant, comes to the surface near the divide.

West of the divide Cretaceous rocks dip southwestward. We have crossed the crest of another anticline, and the rock layers now become younger as we proceed westward into the San Juan Basin. A geologic subprovince in its own right, this basin is about 150 miles across; it extends into Colorado on the north. It has been an important petroleum province since oil was discovered here in the 1940s. Subsiding gradually throughout much of Tertiary time, it equally gradually filled with sedimentary deposits—sand, gravel, clay, and volcanic ash from the San Juan Mountains to the north and the southern tip of the Rockies to the east.

The rock sequence along this stretch of highway is not always straightforward because landslides, mostly with Mesaverde sandstones sliding on Mancos shale, have brought large more or less intact blocks down onto the valley floor.

Between mileposts 95 and 94 the Amargo River, flowing westward, has cut deeply incised "gooseneck" meanders in the Mancos shale. Such meanders have two-stage histories: A meandering pattern is established where the gradient is slight and the stream flows slowly. Then, as uplift and climate change (or both) accelerate stream flow, the river cuts downward, still following its old meanders. Today, much of the erosion in this area is accomplished during sudden summer deluges, or is the work of wind.

On the skyline ridge to the west are all the Cretaceous formations above the Mesaverde group, in ascending order the Lewis shale (forming slopes), Pictured Cliffs sandstone (cliffs), and Fruitland and Animas formations. Above the Animas formation, above a break in the rock record, are some Tertiary rocks. In northwestern New Mexico, the Mesozoic-Cenozoic transition, the time when dinosaurs and many other Cretaceous species became extinct, falls within this break in the record.

Monero is a coal-mining town, with mines in the Menefee formation, the shaly lagoon-deposited middle unit of the Mesaverde group. Coal forms from organic matter—mostly dead plant material—collected in swamps, marshes, and peat bogs, then compacted by pressures of overlying deposits. The coal here is soft bituminous coal rather than the hard anthracite found in eastern United States.

Near Lumberton a sharp ridge with dark soils is one of many dikes that cut north-south through this area. The highway crosses dozens of them in the next 15 miles. The intruding magma baked and hardened the shale on either side of the dikes, and the basalt of the dikes and the baked shale, both harder than surrounding rocks, tend to stand up as sharp ridges. One of the dikes may have fed a sill that caps the mesa to the north.

*The dikes near Lumberton and Dulce form ridges and walls stretching for many miles. Two are exposed in this roadcut, cutting through horizontal layers of Tertiary rock.*

Sedimentary rocks along our route dip westward into the San Juan Basin, so those near the highway are progressively younger and younger. We've come through the Mancos shale and Mesaverde group and also, near Lumberton, through the Lewis shale. The Lewis shale looks pretty much like the Mancos shale, but is younger. Like the Mancos, it is marine, deposited in a Cretaceous sea. Total thickness of Cretaceous sedimentary rocks here is more than 8000 feet.

From Dulce, US 64 turns south into a narrow canyon walled by the Lewis shale, capped by the Pictured Cliffs sandstone. Hollows and small caves are common in this sandstone, and in places wind and water have carved fanciful hoodoos. Watch for crossbedding, for channels cut and then refilled, and for discontinuous layers of more resistant rock. The Pictured Cliffs sandstone marks the shoreline of the late Cretaceous sea, the sands of beaches and nearshore waters. Some shale layers show signs of crumpling that took place when they were soft mud deposits.

As the route curves westward once more, following Vaqueros Canyon, we come on still younger rocks: the Animas formation, which spans the Cretaceous-Tertiary boundary, and the pink-tinted San Jose formation, both made up of many layers of sandstone, shale, and conglomerate. In places the San Jose formation erodes into weird mushroom rocks, largely the work of wind and wind-driven rain. We also encounter the southern end of the dike swarm, with the array of north-south dike ridges cutting through Tertiary rock units. The dikes are one to ten feet thick; some are double.

*Near La Jara Canyon, wind and rain have sculptured soft pink sandstone of the San Jose Formation. Note "window" (arrow).*

65

*Wells near La Jara Canyon tap gas trapped in Cretaceous rocks of the San Juan Basin.*

Tertiary rocks thicken toward the center of the San Juan Basin. The sediments that fill the basin were in large part derived from the southern tip of the Rockies and the San Juan Mountains to the north, both ranges brand new in Tertiary time. The basin was forming even as the mountains rose, and Tertiary sediments fill it to a depth of about 7000 feet. They are well exposed in canyons draining northwest into the San Juan River. Dry most of the time, these canyons see sudden floods, the "walls of water" of fact and fiction, powerful erosive forces in the weak Tertiary sediments. Found in stream terraces along some of the washes are Indian sites and fossil remains of extinct mammals.

We enter La Jara Canyon, a tributary of the San Juan, at milepost 57. Tertiary rocks are also exposed here—irregular ledges and slopes of pinkish sandstone and shale that weather reddish brown. The pinkish color is a clue to rocks of the San Jose formation.

Numerous well sites mark the San Juan Basin gas field. The gas occurs several thousand feet below the surface in the Pictured Cliffs and Mesaverde formations. It is naturally pressurized, so pumping is not necessary. Natural gas is a petroleum product; like oil it forms from decaying animal or plant material. The field is extensive, and there are many wells near the highway and up side canyons. Canyons are favored drilling sites, as wells do not have to be as deep. A roadside exhibit in La Jara Canyon explains the initial processing of natural gas as it comes from the wells.

After leaving La Jara Canyon we drop once more through these sedimentary rocks into Gobernador Canyon, another San Juan tributary. Here again are gas wells against a setting of Tertiary strata.

*Near Farmington, Cretaceous rocks rise sharply, forming cuestas and hogback ridges.*

Shortly before reaching Blanco the highway passes again into rocks of the Animas formation. We've crossed the axis of the San Juan Basin; now rocks become older westward. The shift in dip direction of the sedimentary layers is barely noticeable.

Near Blanco US 64 crosses the San Juan River, a major tributary of the Colorado River, fed from headwaters in the San Juan Mountains in Colorado. Between Blanco and Shiprock the highway travels on Pleistocene terraces bordering the river. Bluffs south of the river expose the Animas formation.

The valley of the San Juan River, well watered, is a garden spot for corn, beans, alfalfa, fruits, and other crops. Salmon Ruins, between mileposts 10 and 9, show that 11th to 13th century Indian peoples also appreciated the fertility of the valley.

The rock strata continue to rise as we proceed westward. Just east of Farmington we come into Cretaceous strata again—the lower part of the Animas formation, then west of Fruitland the Lewis shale and the sandstone-shale-coal-sandstone sequence of the Mesaverde group—almost the mirror image of exposures on the other side of the San Juan Basin. Bluffs across the river expose these Cretaceous rocks well.

Farmington lies on old river terraces developed in Pleistocene time when the river was swollen with many times its present discharge.

Small abandoned coal mines can be seen in some of the rocks north of the highway. The Four Corners Power Plant near Farmington burns Cretaceous coal from larger mines. Smoke plumes from this plant have been blamed for dulling the once crystal air of the Four Corners region, and for endangering the health of those who live downwind from its smokestacks.

67

The dip of the sedimentary rocks steepens west of Fruitland. They turn abruptly upward to form sharp hogback ridges that poke through terrace surfaces on both sides of the river. There are more than 9000 feet of Mesozoic rocks here, so the hogbacks edge the road for some distance, with units becoming older westward.

Shiprock is discussed under US 666 Gallup—Colorado, the next section.

*Along the San Juan River, bluffs of Mancos shale show that the Cretaceous rocks have leveled out.*

# US 666
# Gallup—Colorado
### 111 mi./178 km.

From Gallup to Shiprock this route is on Cretaceous sedimentary rock—mostly the monotonous gray marine shales of the Menefee and Mancos formations. Despite the monotony, there are many interesting geologic features.

Leaving Gallup, the highway climbs onto the top of a mesa surfaced with rocks of the Menefee formation, part of the Mesaverde group. This unit consists of light-colored, often crossbedded, fairly resistant sandstone layers that alternate with dark, slope-forming shale, and with seams of coal near the bottom and top of the formation. With the alternating hard and soft layers we get the ledge-slope-ledge type of weathering so common in the Colorado Plateau region, as well as many picturesque erosional forms: buttes, finger-like pinnacles, and mushroom rocks.

West of the highway, older rocks rise to the surface in a long, deeply canyoned escarpment. These rocks dip eastward as the eastern slope of the great anticline of Arizona's Defiance Plateau.

The Chuska Mountains north of this escarpment also straddle the state line—their northern end is in Arizona. They, however, are capped with thick Tertiary sandstone, the Chuska formation—an isolated body of rock that occurs nowhere else in either New Mexico or Arizona. The unit *may* be related to the Bidahochi formation of Arizona, thought to have been deposited in a large lake. The whole east side of these mountains is covered with landslide deposits—boulder-studded debris on a humpy slope that contrasts with the precisely defined slopes and mesas east of the highway.

Shiprock, a desert landmark, is a textbook example of a volcanic neck with dikes radiating from it.

Oil wells west of the highway pump oil trapped in an anticline in Pennsylvanian rocks.

Uranium occurs in Jurassic shale at the north end of the Chuska Mountains (see map with I-40 Arizona-Grants). Mines are not active at present.

Landslides along the east side of the Chuska Mountains contain boulders as big as houses.

A hard layer of Cretaceous rock, the Dakota sandstone, forms a prominent ridge marking the edge of the San Juan Basin.

From the rest stop near milepost 79 there are good views of nearby dikes. Note their crisscrossing joints.

"Dust devils" whirl sand and silt skyward, helping to further remove the thin soil cover of the San Juan Basin.

Sand dunes near Naschiti exemplify desert processes that are becoming more and more widespread as cattle and sheep destroy vegetation that formerly held down the soil.

Sandstones of the Menefee formation erode into mesas, buttes, and mushroom rocks along US 666. The highway remains on Cretaceous rocks from Gallup to Colorado.

COLORADO

Shiprock

Newcomb

Sheep Springs

Gallup

ARIZONA

San Juan R.

Mancos shale

sandstone?

Chaco R.

Chuska Mtns

sed. rocks

landslides

Mancos-Dakota fm.

Menefee fm.
(Mesaverde gp.)

**US 666**
**GALLUP—COLORADO**

70

The barren gray desert east of US 666 is part of the San Juan Basin, a major oil province. Cretaceous and Tertiary sedimentary rocks bow down there into a large, shallow sag 100 miles across—well known now because of many oil wells drilled there. Oil and gas were discovered in the northern San Juan Basin in the 1940s, with production continuing to the present.

The area around Naschiti is a lesson in desertification at the hand of man. Once richly covered with grass and other vegetation, it has suffered from overgrazing to the point that the grasslands are now becoming sand dunes. Wind is a major factor in erosion here. It picks up the fine silt of the shale layers, leaving behind the sand, which accumulates as dunes. The sandstone caps of the mesas are more resistant, but cliffs and bluffs retreat as they are undermined by the wind. What little moisture falls either sinks into the sandstone of the mesa tops, or erodes the soft silt of the slopes.

North of Newcomb the route enters a land of volcanic necks, dikes, and ubiquitous prairie dog burrows. Bennett Peak west of the highway at milepost 64, and Ford Butte to the east, are small volcanic necks, the lava-filled conduits of volcanoes that once decorated the landscape here. Dikes radiate outward from them, forming cross-country ridges that can be traced in some cases for many miles. There are more volcanic necks to the north—among them the famous Shiprock.

Near milepost 68 the highway leaves the Menefee formation; from here on it travels on Mancos shale, a dark gray marine shale also of Cretaceous age but somewhat older than the Menefee. The springs at Sand Springs, just east of the highway at milepost 62, nourish an oasis of greenery in this otherwise arid land. Groundwater is very near the surface here.

To the east beginning at milepost 71, the long ridge of Hogback Mountain marks one flank of a long, narrow anticline. The hogback itself is formed of steeply tilted Cliffhouse sandstone, part of the Mesaverde group.

To the west, oil wells dot the crest of the anticline—a clue to one type of geologic structure that forms a trap for oil. Since oil floats on water, it tends to work its way upward through porous, water-saturated rock—in this case sandstone—until it reaches an impermeable layer. Unless the sandstone is perfectly horizontal it will then migrate up-dip as far as it can. Here, upward-migrating oil is trapped by impermeable rock layers arching over the crest of the anticline. There it remains, in a sort of upside-down pool. Anticlines on the surface commonly reflect anticlines at depth, so wells such as these perch on top of them. Oil may also be trapped along faults where a permeable layer happens to abut impermeable rock, or in areas

*Shiprock, New Mexico's most famous volcanic neck, towers 1100 feet above the desert. Its massive rock is midway between intrusive and volcanic rock.*

where sandstone or other porous rock layers thin and "pinch out." Traps of all these types exist in the San Juan Basin.

At milepost 79 the highway passes close to the chubby finger of another volcanic neck, with dikes radiating from it. A dike from yet another neck just north of milepost 80 has been put to use as a shelter for a Navajo hogan and its modern equivalent, a house trailer. Note the brick-like jointing of the igneous rock, which is only about two feet thick. The criss-crossing joint patterns of the volcanic necks are visible from the rest stop nearby. Joints in both dikes and necks were caused by shrinking when the rock cooled.

Farther north, Shiprock rises 1100 feet above the surrounding desert. This photogenic peak figures in many geology textbooks as an example of a volcanic neck, complete with radiating dikes that originated as molten rock flowed into cracks radiating from the center of the volcano. Small pinnacles near the main peak are smaller necks of auxiliary volcanic vents. Vertical cooling cracks on Shiprock produce an irregular columnar jointing.

The Four Corners Power Plant east of Shiprock, just visible to the east from milepost 90, burns coal from the nearby Mesaverde group, coal that was laid down as dying plant material in coastal swamps and marshes of Cretaceous time. The plant burns seven million tons a year, and generates 2085 megawatts of electricity, tying in with the power network that serves the entire Southwest. Coal from the San

Juan River area was used by local Indians long before its discovery by white men.

The sea in which the Mancos shale and Mesaverde group were deposited came from the east, sweeping across the flat central part of the continent and well into Arizona. At the end of Cretaceous time the sea withdrew from the continent. All the younger sedimentary rocks in this area, including thick Tertiary deposits that fill the center of the San Juan Basin, are continental, deposited on land rather than in the sea.

The town of Shiprock lies among silty hills of Mancos shale, in the well-watered valley of the San Juan River, one of the main tributaries of the Colorado River. To the north rise the steep flanks of Mesa Verde, capped with thick sandstone layers of the Mesaverde group, among them the Cliffhouse sandstone that we saw in Hogback Mountain. These sandstone layers represent beaches and bars along the edge of the shallow Cretaceous sea. High mountains visible farther north over Mesa Verde are the San Juans, headwaters of both the San Juan River and the Rio Grande.

North of the town of Shiprock the highway climbs through yellow bluffs of Mancos shale onto a broad tableland surfaced with that formation's gray silt. The ground is coated thinly with pebbly desert pavement. To the north, long fingers of Mesa Verde reach toward the valley of the San Juan, their steep gray slopes composed of Mancos shale. Isolated mesas, buttes, and pinnacles, stages in the weathering and erosion of the mesa, stand out at the southern end of many of the fingers. Note the beautifully precise sculpturing on the steep slopes, the treelike branching of gullies, the sharpness of ridges—all features common in arid climates. The main mass of Mesa Verde, as well as Mesa Verde National Park, is in Colorado. West of Mesa Verde is a mountain known as the Sleeping Ute, also in Colorado. Its highest peak forms the Ute's folded arms, while paired volcanic necks, emerging from beneath a forest blanket, form his toes. The mountain is a Tertiary intrusion of a type known as a laccolith, whose intruding magma domed up overlying sedimentary rocks. Though the sedimentary rocks have since eroded from the mountain crest, they show up in rings around its base.

*Tumbled blocks of sandstone litter shaly slopes of Mesaverde group mudstone. Both sandstone and mudstone are products of the last sea to advance across this part of the continent.*

# NM 602, NM 53
# Gallup—Grants
## via El Morro National Monument
## 99 mi./159 km.

For map and section, see I-40 Arizona—Grants, pg. 42

This route circles south of the Zuni Mountains, past El Morro National Monument and El Malpais lava flows, now also the center of attraction in a new national monument.

South of Gallup the highway rises onto a mesa capped with sandstone deposited along the shore of a Cretaceous sea, part of the Mesaverde group. As the shoreline fluctuated with changes in the level of the land, sand (much of it crossbedded) collected on beaches and offshore bars, while gray mud, which would later become shale, accumulated in quiet lagoons. Dying plant material, which would turn into coal, collected in near-shore swamps. Roadcuts expose some of the dark shale and coal seams, the latter too thin and too discontinuous to mine. In its offshore bars, beaches, lagoons, and swamps this area may in Cretaceous time have resembled the present Georgia and Carolina coasts.

Modern valleys along the highway route are filled with fine, silty stream deposits, which in turn are cut by deep gullies that result from removal of plant cover by grazing.

South of milepost 133 the highest parts of the pinon- and juniper-covered tableland are surfaced with remnants of a younger rock—the Bidahochi formation—deposited in Tertiary time in a large lake that extended from here well into eastern Arizona. The lake deposits are flat-lying, whereas the Cretaceous rocks dip eastward into a syncline between the highway and the Zuni Mountains, just out of sight to the east. Clearly, structural changes that created this syncline and the Zuni Mountains came about after the end of Cretaceous deposition, and before the Bidahochi lake deposits accumulated.

The Zuni Mountains are a slipper-shaped, fault-edged dome cored with Precambrian granite and metamorphic rocks. Cretaceous and older sedimentary rocks once arched clear across them. The sedimentary rocks have eroded away from the highest part of the range, however, and now occur only in concentric rings around the mountain core, with harder layers jutting up as tilted, sandstone-capped cuestas and softer ones forming racetrack valleys. The magnitude of the Zuni Mountain uplift can be gauged from the thickness of nearby sedimentary rocks: 4100 feet of Paleozoic, 9500 feet of Mesozoic, and 6600 feet of Cenozoic rocks—20,200 feet in all—have been eroded from the summit of the uplift, exposing the Precambrian core. And the core still reaches summits of 9000 feet or more.

South of the mesa capped with lake deposits, the highway again encounters Cretaceous rocks. The most easily recognized are the Gallup sandstone, a ledge- and cliff-former distinguished by its reddish color, and the Mancos shale, a gray valley-former. Below the Mancos is the Dakota sandstone, the thin but widespread beach-deposited sandstone that marks the bottom of the Cretaceous sequence. The highway rides on soil-covered Dakota sandstone at the junction with NM 53.

There the route turns east, following the racetrack valley of the Mancos shale, with good glimpses of a distinctive reddish or salmon-pink cliff-former, the Zuni sandstone, a Jurassic rock that underlies the Dakota sandstone. Undermined by erosion of another layer of shale below them, pink and cream-colored Zuni sandstone cliffs break down into tumbled blocks of rock. The formation varies in color, and also in resistance to erosion: Where few joints are present, it stands as steep cliffs. Where it is cut by many joints, erosion has carved weird pinnacles and balanced rocks, among them "Los Gigantes," the Giants, east of Ramah.

The town of Ramah lies between two of the cuestas that circle the Zuni Mountains—a Cretaceous cuesta to the south and a Jurassic one

*A lava tunnel forms as molten lava flows out from under its own cooling crust. This one was several miles long before portions of its roof collapsed.*

to the north. North of the town is a narrow gorge carved in pink-striped Zuni sandstone.

The highway continues southeastward along the racetrack valley between these ridges. Between Los Gigantes and El Morro it crosses a low divide into another of the concentric valleys, this one floored with Triassic shale and a grass-covered lava flow. At El Morro National Monument the Zuni sandstone cuesta that separates the two valleys juts northward as bold, weather-streaked cliffs on which passing travelers—dating back to Spanish times—have inscribed their names.

A few miles east of El Morro, beyond another headland of Zuni sandstone, the highway converges with the southeast end of the Zuni Mountains. There it enters a volcanic realm, signaled by hummocky, sage-covered lava flows. A few islands of pale gray Permian limestone jut through the lava; this rock is quarried near milepost 51.

Between mileposts 60 and 61, NM 53 crosses the Continental Divide. The divide is obscure here; nevertheless, it marks the line

*Much of the many-layered ice in Ice Cave may be thousands of years old. In this lava tunnel, the ice rarely sees the sun.*

76

between west-flowing Colorado River drainage and east-flowing Rio Grande drainage. Elevation at the divide is 7882 feet.

Just east of the divide, the small mountains south of the highway are cinder cones. Near them is Ice Cave, one of several lava-tube caves in this area,formed when streams of still-molten lava flowed out from underneath their own cooling, hardening crust. Now part of El Malpais National Monument, Ice Cave is worth a visit, as is Bandera Crater nearby. Cold air, heavier than warm air, sinks into Ice Cave in winter and is trapped there, its chill maintained even through the summer by the insulating properties of lava. Water seeping into the cave—probably over thousands of years—has built up a thick deposit of blue-green ice.

*Bandera Crater, a large cindercone, is the source of the Bandera Lava Flow.*

The trail to Bandera Crater leads past the collapsed canyon that is the head of the original lava tunnel. Here, eruption of lava may have begun soon after the first cinder eruptions. Some basalt magmas contain a high proportion of steam, and at the start of an eruption, when pressures are released, the steam rises into upper parts of the underground volcanic conduits as froth rises when a bottle of beer is uncapped. As the frothy lava is released, pressures are further re-duced, and more froth forms, bursting upward and then falling to the ground to form a cinder cone. Eventually, as the froth is expended, less gaseous magma may flow out on the surface, escaping usually from the base of the loosely consolidated cinder cone. Numerous bubble holes show that even this lava still contains some gas.

The Bandera flow is composed of jagged, broken lava produced when the surface of the flow cooled and hardened while underlying lava was still in motion. In places molten lava splashed up out of minor vents to form little spatter cones like the one near the Bandera Crater trail. On the slopes of the main crater are many volcanic bombs tossed out in semi-molten form by the eruption. Some are football-shaped from their spinning journey through the air.

East of Bandera Crater the highway runs close to the contact between reddish Precambrian granite—the core of the Zuni Mountains—and the rough lava of the Bandera flow, in most places thickly forested. In all, the flow is about 25 miles long and 10 miles wide.

As the route rounds the southeastern tip of the mountains there are good views of their red granite core, which forms a sort of saddle between flat-lying and east-dipping Paleozoic sedimentary rocks. The highway crosses some of the Paleozoic strata. Cretaceous rocks wall lava-capped Cebolleta Mesa to the east. Large alluvial fans emerge from canyons on this side of the valley.

Soon Mt. Taylor, an eroded composite volcano, comes into view to the north. Mt. Taylor and the lava-capped mesa on which it rests are Tertiary volcanic features, older than the cinder cones and lava flows we've just seen.

At San Rafael, Paleozoic strata dip northward off the northeast side of the Zunis, completing their circuit of the Zuni Mountain dome. Because lava flows dam the valley, drainage is poor near San Rafael, and white alkaline salts collect on soil surfaces.

# III
# The Rift and the Rockies

Cutting southward across part of Colorado, all of New Mexico, and much of the Mexican state of Chihuahua, the Rio Grande Rift is a major break in the Earth's crust. In it, a long sliver of crust has dropped thousands of feet between two irregular but very deep fault zones. Fault movements began about 30 million years ago; they continue to the present day. Portions of the rift see frequent minor earthquakes, and young alluvial fans along the edges of the rift are in places marked by fault scarps obviously even younger than they are.

How does the Rio Grande Rift relate to what we know of plate tectonics in this area?

The rift is made up of a series of staggered basins, a long north-south string of them. It is caused by tension, a pulling apart of the Earth's crust, and the moving away of hot, plastic mantle material from beneath the down-dropped wedges. (Actually, this is putting the cart before the horse! Movements of mantle material probably came first and caused the rift.) In places, these wedges have dropped some 26,000 feet—about 5 miles. If it weren't for the immense amounts of gravel and sand and lava and volcanic ash that fill the rift, a narrow seaway comparable with the Red Sea or the Gulf of California, or a scorching below-sea-level valley like that of the Dead Sea in Israel or the Imperial Valley of southern California would cut New Mexico neatly in half.

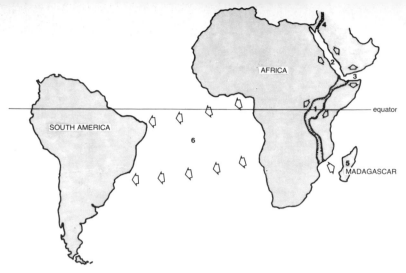

*Africa furnishes a living lesson in the growth and development of rift valleys. Its Great Rift (1), similar in many ways to the Rio Grand Rift, splits and deepens to form the Red Sea (2) and the Gulf of Aden (3). Part continues north as the Dead Sea Rift (4). At the southern end of the Great Rift Madagascar (5) has broken away from Africa. Another rift system west of Africa has widened into the South Atlantic Basin (6).*

Along the edges of the rift, volcanism is common, and has been since mid-Tertiary time. Once-towering composite volcanoes and eroded volcanic necks are mostly of Oligocene and Miocene age, as are silicic lava and ash flows that make up the walls of former calderas. Younger basalt lava flows and cinder cones give evidence that the faults bordering the Rio Grande Rift reach down to the Earth's mantle, the ultimate source of most basaltic rocks.

The Rio Grande Rift starts as a narrow sliver in central Colorado. It broadens southward, until near Albuquerque it is about 30 miles wide. Farther south it gets still wider, and includes not only the valley of the Rio Grande, but Jornada del Muerto and the Tularosa Valley to the east—both of them bordered by faults of the same age as the rest of the rift.

When first established, the rift appeared as a number of closed basins—short troughs that did not drain or even intercommunicate. Gradually filling with sediments washed in from adjacent ranges, as well as with lava flows and volcanic ash from nearby volcanoes, the string of basins slowly joined up, affording the young Rio Grande a route southward within its confines.

*Many of New Mexico's early pueblos were partly destroyed by earthquakes centered along the Rio Grande Rift.* Hake C. Avon photo, Earthquake Information Bulletin 56, courtesy of U.S. Geological Survey.

In southern New Mexico the rift and its bordering ranges blend with valleys of the Basin and Range region, there being no clear geologic distinction between them. Both the Rio Grande Rift and the Basin and Range region came about in response to the same great pulling-apart tensions in the Earth's crust; in both, these tensions set a pattern of tilted ranges and intervening valleys built and deepened from mid-Tertiary time on. This chapter therefore includes the Basin and Range parts of southwestern New Mexico.

In the northern part of the state the rift splits the southern end of the Southern Rocky Mountains. Since here the development of the mountains is intimately tied in with development of the rift, the two-pronged southern end of the Rockies is discussed in this chapter also.

The location of the Rio Grande Rift may have been controlled by some much older geologic patterns—an older zone of weakness or possibly an earlier rift in the Earth's crust. The same can be said of the present Southern Rocky Mountains, which lie in almost the same position as the Ancestral Rocky Mountains established in Pennsylvanian time.

The Rio Grande Valley is not a valley in the usual sense of the word: It was not cut by a river, and it doesn't branch upstream the way most river valleys do. The Rio Grande merely found and followed the pre-established, partly filled-in rift trough. The river, however, *has* markedly influenced the present shape of this trough. In alternating cycles of deposition and downcutting, the river filled the fault-edged trough with sand and gravel brought in from the adjacent mountains and from upstream, and then cut into it, alternately building and trenching, building and trenching, to create a series of steplike terraces along its margins.

Geologists lump these Tertiary terrace deposits into the Santa Fe group. This unit varies from place to place, as you can imagine, because of differing sources of sediment and different kinds and amounts of lava flows and volcanic ash flowing or blowing into it. By and large the sediments are soft and easily eroded. In many places they have been tilted by continuing movement along rift valley faults. In areas where they contain large amounts of volcanic ash, they erode into badlands.

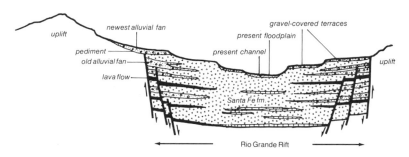

*In cross-section, the Rio Grande Rift is filled with sand, silt, gravel, clay, lava, and ash flows of the Santa Fe group, thinly covered with younger terrace deposits.*

Both geologic changes and cyclic changes of climate influenced the history of the Rio Grande in this area. At one time the river flowed on the surface of the Santa Fe group, which blended with eroded pediments at the bases of the mountains. Later, probably because of climatic changes, it cut down through this surface. During times of increased river discharge, corresponding to glacial episodes far to the north, major valleys were shaped. During drier episodes corresponding to interglacial stages, both erosion and sedimentation increased, largely because the climate was arid and there was much less vegetation to hold down soils. The same pattern can be recognized today: During times of low discharge the Rio Grande deposits gravel bars and sand bars; with increased river discharge it cuts into its established floodplain.

# I-10
# Deming—Texas border
### 79 mi./127 km.

On the eastern outskirts of Deming the highway crosses the Mimbres River—usually dry, but with headwaters to the north in the Mimbres Mountains. When it does flow, the river carries water southeastward around the Florida Mountains, and then sinks into the porous gravels of the desert.

Rockhound State Park, 12 miles southeast of Deming near the north end of the Little Florida Mountains, encourages rock collecting. Agates and geodes are found among the volcanic rocks there. Mines on the northeast slope of the Little Florida Mountains produced manganese, used to harden steel and to manufacture flashlight batteries.

The Florida Mountains contain Precambrian granite, Paleozoic sedimentary rocks, Cretaceous redbeds (red sedimentary rocks), and Tertiary volcanic rocks—particularly a great thickness of volcanic breccia, broken rock debris produced by huge volcanic explosions. The Cretaceous sedimentary rocks, quite thick here, show us that this area was near the fluctuating shoreline of a Cretaceous sea.

To the north from mileposts 90-95 are more Tertiary volcanic rocks. Cooks Peak (8404 feet) is a small intrusion that forced its way up through the sedimentary rocks, some of which now remain in a ring around it. The westernmost of the faults that define the Rio Grande Rift runs through the Cooks Range, so from here eastward we are technically within the rift, even though the land remains the same topographically.

The "Hatch cutoff" passes north of volcanic tuff that makes up much of the Sierra de las Uvas. With US 85 and I-10, it forms a scenic and geologic loop trip.

Scenic US 85 stays close to the Rio Grande, passing below high river terraces and through a narrow gorge where the river cuts through Paleozoic sedimentary rocks and Tertiary intrusions.

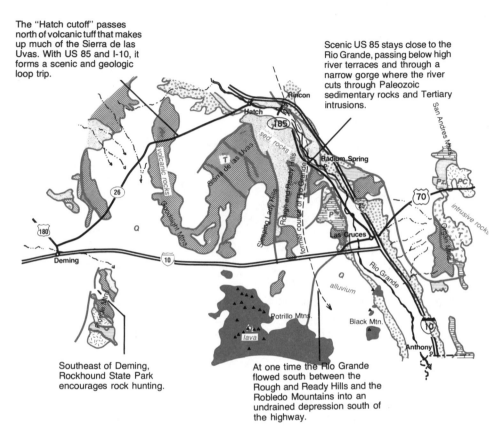

Southeast of Deming, Rockhound State Park encourages rock hunting.

At one time the Rio Grande flowed south between the Rough and Ready Hills and the Robledo Mountains into an undrained depression south of the highway.

## Interstate 10
## DEMING—TEXAS

## NM 26
## DEMING—HATCH

## US 85
## LAS CRUCES—RINCON

Mines near Cooks Peak, most of them now closed, produced silver, lead, zinc, and fluorite. Fluorite occurs in veins associated with the lead and zinc ores. As a source of fluorine, it has a wide variety of industrial uses: in manufacture of glass and glazes, hydrofluoric acid used in metallurgy, teflon, aerosols, freon for refrigerators, and toothpaste.

The Potrillo Mountains to the south between mileposts 110 and 115 represent a broad but low volcanic field, a cluster of volcanic cones and lava flows. There are about 150 cinder cones in this volcanic field, which is no more than 150,000 years old, and lava flows cover hundreds of square miles. Several low craters on the fringes of the volcanic field result from steam explosions that occurred when hot lava met up with groundwater. To the north, the Sleeping Lady hills and Rough and Ready Hills owe their origin to Tertiary volcanism.

Between mileposts 120 and 129, the highway crosses a shallow, wind-scoured depression that contains pebbles not derived from adjacent ranges. The pebbles were brought here by the Rio Grande at a time when it flowed directly southward from Hatch. Tertiary volcanism may have played a part in deflecting it southeastward to its present channel.

Approaching the Rio Grande Valley near Las Cruces, we can see that most of the town is built on the floodplain and terraces of the Rio Grande. The view to the east includes the Organ Mountains, named for their light-colored craggy outcrops of vertically jointed Tertiary granite. Tilted blocks of stratified rock—mostly Paleozoic—make up the southern part of these mountains. The highway drops down over several terrace levels to the river floodplain, then climbs through their counterparts on the east side of the river. Manmade revetments

*Tertiary granite, eroded along vertical joints, forms the "organ pipes" of the Organ Mountains near Las Cruces.*

near the highway prevent storm runoff from flooding the city and valuable bottomlands along the Rio Grande.

From Las Cruces southward to the Texas border, Interstate 10 continues down the east side of the valley of the Rio Grande, with the Organ Mountains to the east and flat, almost level desert punctuated with a few small volcanoes to the west. Here the Rio Grande Valley is only a small part of the Rio Grande Rift, which also includes the Tularosa Valley to the east.

Soils that are hardly more than sand blow up out of the Rio Grande's bed or off adjacent floodplains and terraces. Spring plowing unfortunately coincides with New Mexico's spring winds, to the detriment of the sandy soils. In places, drifting dunes have developed here, many of them built up around sand-catching vegetation. In the mild, almost frost-free climate, and with the Rio Grande to supply irrigation water, the floodplain is farmed intensively; alfalfa and cotton fields alternate with pecan groves and fruit orchards.

Near the Texas border, steeply dipping Paleozoic limestones trace chevrons on ridges of the Hueco Mountains. These marine limestones, deposited in Pennsylvanian and Permian seas, extend south and southeast into Texas as well as east and northeast across southern New Mexico.

# I-25
# Las Vegas—Santa Fe
**64 mi./103 km.**

Still following the route of the Santa Fe Trail, Interstate 25 leaves Las Vegas to pass through a gap in the lines of hogbacks that edge the southern tip of the Sangre de Cristo Range. Magnificent roadcuts expose the tilted rocks of these hogbacks: the Cretaceous Dakota sandstone, many-hued Jurassic Morrison formation, and finally Triassic red sandstone and shale. Near the junction with US 84, steeply sloping Triassic, Permian, and Pennsylvanian sedimentary rocks introduce older rock layers that then border the highway most of the way to Santa Fe.

Triassic and Permian strata include, in the order in which you'll encounter them, light gray and tan Santa Rosa sandstone, purplish red Bernal formation, gray San Andres formation (especially well exposed in the roadcut between mileposts 332 and 331, where some nice little faults offset the layered limestones), Glorieta sandstone, and a thick sequence of Yeso and Sangre de Cristo formation "redbeds"—brick red siltstone and sandstone. The Yeso formation includes many narrow veins of gypsum, in the form of the mineral selenite.

Together, these rocks tell a tale:

1. Uplift of mountains in Pennsylvanian time, and their gradual wearing down, providing materials of the Sangre de Cristo formation;

2. Eroding of the land and depositing of near-shore and shallow-water sandstone and mudstone of the Yeso formation;

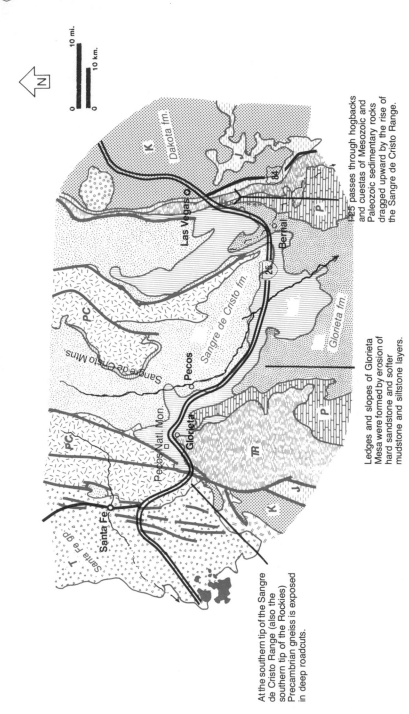

Ledges and slopes of Glorieta Mesa were formed by erosion of hard sandstone and softer mudstone and siltstone layers.

I-25 passes through hogbacks and cuestas of Mesozoic and Paleozoic sedimentary rocks dragged upward by the rise of the Sangre de Cristo Range.

At the southern tip of the Sangre de Cristo Range (also the southern tip of the Rockies) Precambrian gneiss is exposed in deep roadcuts.

**Interstate 25**
**LAS VEGAS—SANTA FE**

3. Development of beaches—the future Glorieta sandstone;

4. More subsidence, with a shallow, warm sea coming in over what had been land, depositing the marine San Andres limestone, largely a product of animal and plant shells;

5. Restriction of the sea and its gradual drying, probably in a desert climate similar to that near the Red Sea today, with deposits of fine siltstone, salt, and gypsum of the Bernal formation;

6. An interval of erosion now represented by an unconformity at the top of the Bernal formation;

7. Depositing of river sediments of the Santa Rosa sandstone.

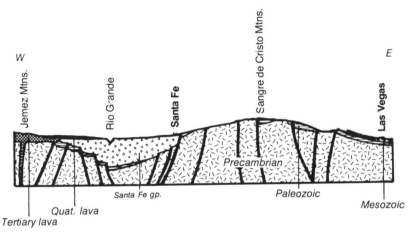

*Section across the Sangre de Cristo Mountains north of I-25 Las Vegas – Santa Fe.*

All these sedimentary layers level out as the highway curves westward around the southern tip of the mountains, where the upper part of the Permian sequence appears as horizontal layers on the escarpment of Glorieta Mesa, straight ahead at milepost 325. Much of the escarpment is heavily vegetated; west of milepost 311, however, the formations can be distinguished. Hard sandstone and limestone layers form ledges and cliffs, while soft mudstone and siltstone layers weather into slopes. The red-brown color of many of the rock units is caused by tiny grains of hematite, an iron oxide that we normally think of as rust. Iron oxides in different amounts impart many shades of pink and yellow to sedimentary rock; they come originally from iron-bearing minerals like biotite and hornblende.

As the highway climbs toward Glorieta Pass, the escarpment on which these rocks are exposed curves with it. Near Pecos it forms a backdrop for Pecos National Monument, where ruins of a mission

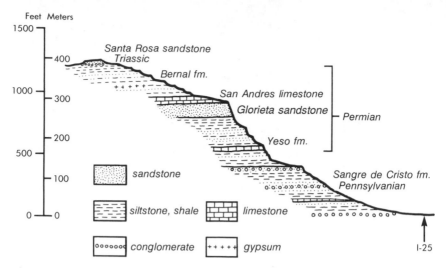

*Stratigraphic diagram of rocks exposed on Glorieta Mesa near Pecos National Monument.*

church and Indian pueblo are built of the very rocks that surround them. Even thin slabs of clear gypsum from the Bernal formation were put to use—as windowpanes!

Glorieta Pass (7432 feet), a few miles west of Pecos, was crossed by plodding horses and footmen of Coronado's expedition in 1540-1541. After 1821 it became a part of the Santa Fe Trail. Later, the railroad and highway followed the same route.

West of the pass the sedimentary formations dip southward off the range. Some of them appear—steeply tilted—in the big roadcut at milepost 296. The Sangre de Cristo formation makes up the hills north of the highway; behind them rise Precambrian rocks that form the core of the mountains. Faulted up on both sides, this range is complex and interesting; roads lead into it from Pecos and Santa Fe, and there are many trails into its delightful canyons. The highest summit, Truchas Peak, is 13,102 feet in elevation.

Watch roadcuts for dikes. From milepost 292, Precambrian rocks of the main mountain mass extend to the highway: very hard, very resistant, greenish black gneiss patterned with veins and dikes. Between here and Santa Fe and for some distance north of Santa Fe a broad band of this gneiss stands as the west side of the range, rising well above the down-faulted Rio Grande Rift.

Across the valley to the west lie the Jemez Mountains, remnant of a composite volcano that a million years ago collapsed into its own magma chamber. To the southeast, behind scattered hills of the Ortiz Mountains, is the Sandia Range. Broad slopes between here and the

Ortiz Mountains are known to geologists as the Ortiz surface, a partly eroded, partly deposited Tertiary surface that existed before the Rio Grande became a through-flowing, down-cutting river.

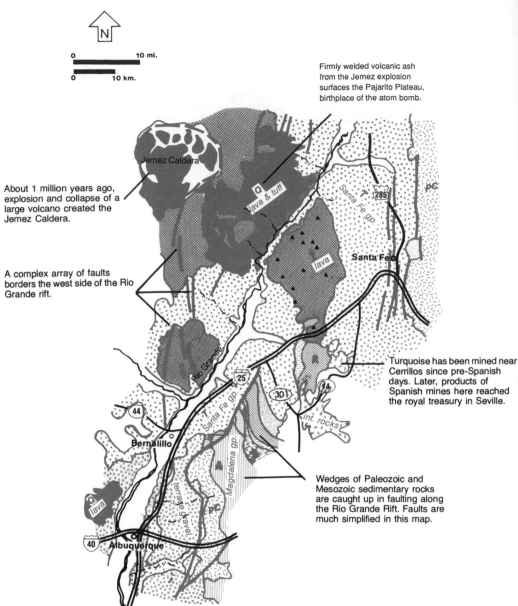

Firmly welded volcanic ash from the Jemez explosion surfaces the Pajarito Plateau, birthplace of the atom bomb.

About 1 million years ago, explosion and collapse of a large volcano created the Jemez Caldera.

A complex array of faults borders the west side of the Rio Grande rift.

Turquoise has been mined near Cerrillos since pre-Spanish days. Later, products of Spanish mines here reached the royal treasury in Seville.

Wedges of Paleozoic and Mesozoic sedimentary rocks are caught up in faulting along the Rio Grande Rift. Faults are much simplified in this map.

**Interstate 25**
**SANTA FE—ALBUQUERQUE**

# I-25
# Santa Fe—Albuquerque
### 56 mi./90 km.

Southwest of Santa Fe, Interstate 25 crosses the broad, gentle slope eroded into the receding mountain base of the Ortiz Mountains to the south. This Ortiz erosion surface, about 400 feet above the present Rio Grande, can be recognized through much of northern New Mexico. Here as elsewhere it is covered with a thin veneer of Quaternary gravel, the upper part of the Santa Fe group.

South of the Sandoval County line (milepost 270) the highway rises onto a basalt surface studded with small, low, somewhat obscure volcanoes. Here the Ortiz surface merges with the lava surface, which also wears a thin veneer of gravel. As you leave the basalt, you'll see some reddish and pinkish sandstone and mudstone below the basalt right at the side of the highway. These weak, easily eroded rocks, part of the Tertiary Santa Fe group, are well exposed between mileposts 264 and 262. At milepost 263, where the highway crosses the Rio Cerillos, a gypsum mine can be seen upstream. Another is upstream from milepost 253. The gypsum, which occurs in Jurassic rocks, is used in manufacturing cement, wallboard, and plaster. Mica is mined near here as well.

The Ortiz Mountains, as well as the San Pedro Mountains farther south, consist of clusters of small Tertiary intrusions—stocks, sills, and small laccoliths—which pushed up through the fault zone along the east side of the Rio Grande Rift. Among these intrusive rocks are slivers of Triassic and Jurassic sedimentary rock and numerous dikes, as well as some volcanic rocks.

Hot, mineral-rich fluids of the intrusions altered some of the surrounding rocks. One of the more interesting products of this alteration was turquoise, which has been mined here since pre-Spanish days, perhaps as early as 900-1100 A.D. Valued by prehistoric inhabitants of the Rio Grande Valley for both ornament and ceremony, turquoise from this area was also a major item of trade: Some of it reached Teotihuacan, the Aztec capital that became Mexico City, and after the Spanish Conquest found its way by treasure galleon to Spain, where it took its place among the crown jewels.

In 1828 gold was discovered in these little ranges—placer gold that could be washed from stream gravels. Production was limited, of course, by the lack of water—a problem in most of New Mexico. Lode gold was found in bedrock here a few years later, inaugurating a gold rush that preceded the more famous California gold rush by more than a decade. Later, lead, silver, and hard anthracite coal, rare in the West, were mined in these dark hills.

The highway gradually converges with the lava-walled gorge of the Rio Grande. This route furnishes some good views of the Rio Grande Rift, the great trough that cuts New Mexico in half. Bound on both east and west by staggered and locally complicated fault zones, the rift forms a broad, somewhat zigzagging trough trending, in this area, northeast to southwest. Within it the Rio Grande has carved its modern channel, well marked by a line of cottonwood trees.

*In profile, the Sandia Mountains show their steep western face (right) carved in Precambrian granite, and their gentler eastern slope, a dip-slope of Paleozoic sedimentary rocks. The Ortiz erosion surface is visible in the foreground.*

Southwest of the Ortiz and San Pedro mountains, beyond a low saddle, is the great dome of the Sandia Mountains. This range is a single east-tilted fault block cored with Precambrian granite and metamorphic rock. Lifted some 11,000 feet above sea level, and

94

26,000 feet (5 miles!) above corresponding granite beneath the Rio Grande Rift, the Sandia Range is surfaced on its relatively smooth, gentle eastern slope with several hundred feet of Pennsylvanian sedimentary rocks, mostly limestone layers of the Madera formation. The steep, craggy western slope, far different in appearance than the eastern slope, is the faulted side of the tilted block, worn back and ornately sculptured by erosion.

Look west from milepost 245 to see several small volcanoes neatly sliced open by the Rio Grande. You can easily see their former central conduits, which jut above the surface as volcanic necks. In this general area, quite a few small volcanoes—the sources of lava flows—line up along north-south fissures.

During its early history the Rio Grande Rift consisted of a series of closed non-draining basins, many of them occupied by lakes. Gradually the basins filled with sediments washed in from adjoining ranges, as well as with lava flows and volcanic ash, deposits now known collectively as the Santa Fe Group. As the basins filled, the Rio Grande established itself as a through-flowing stream. The largest of the former basins, now filled with several vertical miles of gravel, lava, and ash, is the Albuquerque-Belen Basin, with Albuquerque lying at its eastern edge.

Coming into Albuquerque, look across the Rio Grande's inner valley to see the small Albuquerque volcanoes, another string of small cones lined up along a north-south fissure.

Albuquerque's water needs are met by wells tapping the porous, permeable gravel of the valley fill. These in turn get water from precipitation and from underground flow from the Rio Grande and its

*A small tributary has built a triangular white delta out into the channel of a larger stream. Note the "braided" pattern of the large stream, characteristic of watercourses in arid climates.*

tributaries, as well as from unrenewable "fossil" water trapped in these sediments when they were deposited. Manmade levees along the mountain front protect the city from sudden floods originating from cloudburst-style storms over the Sandias. Some of the water caught by the levees sinks into the porous ground, adding to the general supply, rather than flooding through the city and onto the low bottomlands that border the river's present channel and natural levees. Starting with a major flood in 1874—which crested higher than any previously recorded flow and established a new river channel—the low land near the river became progressively more and more swampy and waterlogged and unsuitable for farming. In 1928 a program was established to drain this area and to prevent further flooding, a task carried out in the 1930s and '40s by closing off old irrigation ditches, cutting new drainage ditches, and diverting mountain runoff and rerouting it directly to the main river channel.

*A life-size model of* Pentaceratops *welcomes visitors to the New Mexico Museum of Natural History in Albuquerque.*

## Albuquerque and vicinity

The city of Albuquerque rests on several river terraces on the east side of the Rio Grande, below imposing cliffs of the Sandia Mountains. Beneath the recent gravels that coat them, the terraces are composed of Miocene and Pliocene sediments of the Santa Fe group.

*Granite weathering along joints becomes more and more rounded, eventually forming residual (weathered in place) boulders.*

New Mexico's Museum of Natural History, located in Albuquerque, presents an excellent up-to-date review of the geologic history of the state. Exhibits include bones and reconstructions of dinosaurs and other Mesozoic reptiles, dioramas showing fossil mammals that once roamed this region, a reconstructed volcano, and rock and mineral specimens. Highly recommended!

The terraces on which the city rests represent intervals during which the Rio Grande's ability to deepen its channel was balanced by its ability to deposit sand and gravel—a balance upset periodically by the changing warm-dry and cold-wet climates of Pleistocene time and by uplift and fault movement. The lowest terraces are only a few thousand years old. Some are crossed by low scarps marking recent movement on faults edging the Rio Grande Rift. And some contain early Indian sites with hearths, tools, and in some cases bones.

In 1936 another site of early human occupation was discovered in a cave at the northeast end of the Sandia Range. Three levels of occupation were found here, one on top of the other. In the uppermost or youngest level, pre-Columbian pueblo-style artifacts were found. In the second layer, skillfully made Folsom spear points, known from near Folsom in northeastern New Mexico, were associated with bones of now extinct bison, mammoth, giant sloth, camel, horse, and wolf. The presence of horse bones reminds us that horses developed in the New World, migrated to the Old, and then became extinct in the New World while continuing their evolution in Asia, Europe, and Africa.

They were reintroduced to the Americas by Spanish conquistadors. The lowest level revealed more bones of extinct animals—bison, camel, mastodon, horse, and mammoth—associated with hearths and implements of nomadic hunters, and with a type of spear point that has come to be known as the Sandia point.

The Sandia Mountains, which rise so boldly behind the city, are an east-tilted fault block of 1.4 billion-year-old Precambrian granite topped with 300 million-year-old Pennsylvanian sedimentary rocks. The height of the mountains is emphasized by subsidence of the Rio Grande Rift: Beneath the rift, the same Pennsylvanian sedimentary rocks are nearly 20,000 feet down. If we add to this figure the height of the mountains above Albuquerque—about 5000 feet—we come up with total fault movement of 26,000 feet—about 5 miles. The fault zone along which all this shifting took place lies hidden beneath big coalescing alluvial fans that border the west side of the range.

Certainly the most exciting way to see these rocks is to ride up to the crest of the range on the Sandia Peak Tramway. Or, you can drive up the Sandia Crest highway, which zigzags up the east side of the range to the highest point on the mountain, 10,678 feet in elevation. As seen from the tram, large boulders of Precambrian granite, weathered right in place, cover the irregular slopes at the base of the range, giving way upward to giant crags and pinnacles clearly resulting from weathering along many joints. Above the second tower the tram converges with sedimentary rocks that cap the range, and you can see them, too, at the summit itself—several hundred feet of marine limestone and dark brown shale. They lie on an irregular surface of granite eroded during the many millions of years of late Precambrian and early Paleozoic time. Both limestone and shale layers contain fossil brachiopods, bryozoans, and corals, as well as buttonlike sections of crinoid (sea lily) stems.

From either Sandia Peak or Sandia Crest, the city of Albuquerque is spread out below, its eastern edge crowding the great alluvial fans

*Pennsylvanian limestones form the summit of the Sandia Mountains. Below them (lower left) are vertically-jointed cliffs of Sandia Granite.*

98

at the mouths of mountain canyons. Good eyes can pick out the flood-control embankments and storage reservoirs east of the city. Beyond the city proper, near the far side of the inner valley, the Rio Grande swings lazily in its channel, below the terraces that edge the valley on the other side. On the highest of these terraces, five little dark pimples are the Albuquerque volcanoes, active in Pleistocene time, rising above their dark-edged lava flows. Far beyond them Mt. Taylor is silhouetted against the western skyline. On the north are forested slopes of the Jemez Mountains, remains of a once-towering composite volcano. With the Sangre de Cristo Mountains near Santa Fe, visible to the northeast, they make up the double southern tip of the Rocky Mountains.

Eastward, partly obscured by the trees, is the relatively smooth east slope of the Sandias, a dip slope surfaced with Pennsylvanian sedimentary rocks—mostly limestone. All younger sedimentary layers have been stripped away by erosion. East of the mountain base the Pennsylvanian limestone dives under the Estancia Basin. Younger rocks are exposed on the other side of this basin, whose fine silts attest to a lake that existed there in Pleistocene time. And to the south, beyond the crest of the Sandias, are the Manzano Mountains, a continuation of the tilted fault block structure of the Sandias.

Water for the city of Albuquerque comes largely from wells that penetrate the terrace deposits and alluvial fans, tapping porous layers in the upper part of the Santa Fe group. Much of the water was "deposited" along with the sediments, so that in effect the wells that tap it are "mining" the water—depleting water that cannot be replaced. But there is also a considerable amount of recharge from nearby Sandia canyons, where runoff sinks into the porous gravel of the alluvial apron along the mountain front. Some of the water used by the city also finds its way back into the terrace deposits. So the supply should be adequate for some time despite rapid urban growth.

Historically, Albuquerque has at times suffered from too much water. Both the Rio Grande and the nearby Sandia canyons have a long history of flooding. Early attempts to prevent Rio Grande floods involved heightening and strengthening the natural levees along its banks, and development of parallel storm channels, which also served to drain low, marshy ground on either side of the main channel—now productive bottomland. More recently, construction of upstream dams also helped to contain floodwaters. Flood control embankments along the east edge of the city capture storm runoff from Sandia cloudbursts, and convey the water to holding basins and ultimately to the Rio Grande. This flood control system is designed to withstand even "100-year" floods—designated by the U.S. Geological Survey as likely to occur on an average of once every 100 years.

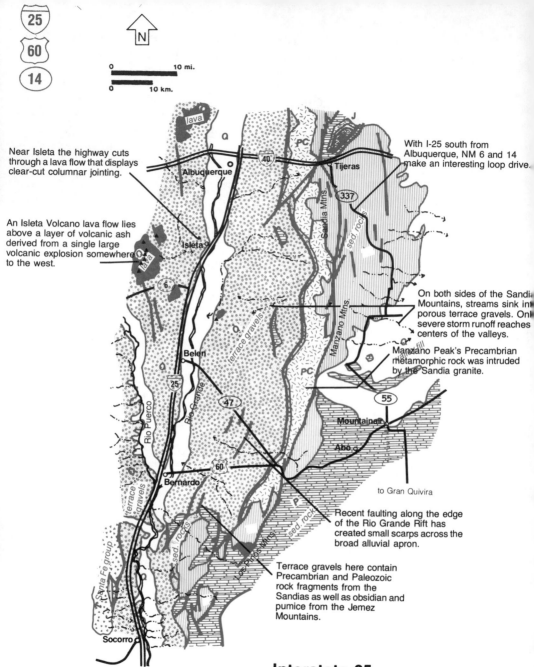

Near Isleta the highway cuts through a lava flow that displays clear-cut columnar jointing.

An Isleta Volcano lava flow lies above a layer of volcanic ash derived from a single large volcanic explosion somewhere to the west.

With I-25 south from Albuquerque, NM 6 and 14 make an interesting loop drive.

On both sides of the Sandia Mountains, streams sink into porous terrace gravels. Only severe storm runoff reaches centers of the valleys.

Manzano Peak's Precambrian metamorphic rock was intruded by the Sandia granite.

Recent faulting along the edge of the Rio Grande Rift has created small scarps across the broad alluvial apron.

Terrace gravels here contain Precambrian and Paleozoic rock fragments from the Sandias as well as obsidian and pumice from the Jemez Mountains.

to Gran Quivira

## Interstate 25
## ALBUQUERQUE—SOCORRO

## SANDIA—MANZANO
## MOUNTAINS LOOP

*Well jointed Sandia Granite forms most of the west side of the Sandia Mountains. The same granite lies thousands of feet down below the Rio Grande Rift.*

# I-25
# Albuquerque—Socorro
## 77 mi./123 km.

The east-tilted block of the Sandia Range, about 20 miles long, dominates the geologic scenery for some distance south of Albuquerque. The major canyon cutting obliquely into the range is Tijeras Canyon, eroded in fractured rock along the Tijeras fault. The highway crosses Tijeras Creek's huge alluvial fan, three miles across, south of the intersection with Interstate 40.

To the west, across the Rio Grande Valley, small pimples on the skyline mark the Albuquerque volcanoes, little cones and basalt lava flows that erupted 150,000 years ago. The volcanoes line up neatly along a north-south fissure. Farther south near Isleta a small shield volcano is the source of five similar flows. The Isleta volcano is superimposed on a ring of tuff produced by an explosive steam eruption that occurred when hot lava came in contact with water.

South of the Sandia Range and forming a continuous uplift with it are the Manzano Mountains, another tilted block of Precambrian rock wearing a cap of east-dipping Pennsylvanian limestone. Man-

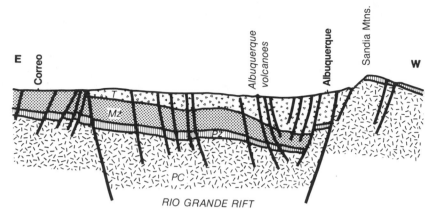

*Section across Rio Grande Rift at Albuquerque.* (Adapted from Hawley and others, 1982)

zano Peak at the south end of the range is dark with Precambrian metamorphic rocks.

Here, as everywhere, the Rio Grande Rift is much deeper than it looks. It is filled in with thousands of feet of alluvial fan, lake, and playa deposits that predate the through drainage of the Rio Grande, and with sediments brought in by the river after through drainage was established.

The highway crosses the Rio Grande north of Isleta. The extra river channel visible from the bridge is a manmade floodway protecting nearby farms and settlements. The river valley here, 30 miles wide, is part of the Albuquerque-Belen Basin, by far the largest of the sediment-filled basins along the rift. Between the bordering ranges and the river there are four terrace levels that represent alternating episodes of valley cutting and partial refilling. Most of the valley fill was deposited during Pliocene and Pleistocene time. Since then the river has carved an inner valley, where it flows through a floodplain dotted with fields and orchards.

Isleta and its nearby pueblo remind us that this broad basin, with its fertile floodplain and abundant water, has provided a bountiful living for its inhabitants for many thousands of years. The original pueblo, on a lava flow remnant that rises like a little island *(isleta)* from the Rio Grande floodplain, was an established town when Coronado arrived in 1540.

In contrast with the fertile floodplain, poor soils of the terraces serve only for grazing, and since the coming of Spanish settlers, overgrazing has unfortunately been the rule. Soils and natural desert vegetation were far more lush before cattle, horses, and sheep were brought into the area. The now ubiquitous but unpalatable creosote bush, which maintains its place by poisoning its plant neighbors, has invaded much of what was described by early Spanish settlers as lush grassland. On several terrace slopes, sand dunes indicate that even the well-adapted creosote and sagebrush cannot completely hold back desertification.

The broad, sloping plain below the Manzano Mountains, visible to the southeast from milepost 209, is broken about halfway up the slope by a fault scarp. Movement on the fault obviously postdates development of the sloping plain, and probably occurred only a few thousand years ago. A good many of the faults that border this basin are still functioning. This is New Mexico's most active seismic region; about 95 percent of the state's hundreds of small recorded earthquakes originate between Albuquerque and Socorro.

Between mileposts 204 and 201 the highway passes close to the small volcano of Sierra Lucero. Lava flowing from this volcano reached as far as the present highway, where it appears in roadcuts. Farther south and west, watch for exposures of the eroded edge of the uppermost terrace and its flat layers of reddish sand, clay, and volcanic ash, part of the Santa Fe group.

West of the junction with US 60 is the dome of Ladrone Peak, an up-faulted wedge of Precambrian rock. Dark volcanic mountains behind it mark the northeast corner of the Datil-Mogollon volcanic highlands of west-central New Mexico.

The interstate crosses the Rio Puerco at milepost 174. This river, often dry and marked with the sinuous patterns of a braided stream, begins north of Grants. When it flows, it carries a heavy sediment load and is a major contributor of sediment to the Rio Grande. It is bordered with terraces of poorly consolidated Tertiary and Quaternary gravel. The valley is famous among geologists as the site of river deposits containing many skeletons of now-extinct mammals; the main fossil sites lie well upstream.

South of the Rio Puerco, the Rio Grande floodplain is broad and marshy. Bird sanctuaries here shelter many species of waterbirds following a flyway between winter feeding grounds and summer nesting sites. The marshes exist because the river is to some extent held in check by converging ranges on either side—the so-called Socorro constriction. South of here, however, the rift broadens as its eastern fault zone swings far east to include the next two valleys: the Jornada del Muerto and Tularosa Valley.

From the rest area at milepost 167, sand dunes can be seen along the edges of nearby terraces. The ultimate result of desertification, dune sands shift and blow about too much to support vegetation. Here

*Gusty winds, a potent force in desert erosion, move sand and silt from the Rio Salado's dry riverbed.*

the sand is derived from the bed of the Rio Salado, a normally dry desert wash that, when it flows, brings a supply of new sand from mountains to the west. Braided patterns common to desert rivers often show up on the streambed. Even a quick glimpse from the bridge shows the sorting of sand, pebbles, and cobbles that is one of the clues to deposition by water.

Contributing to the maintenance of the marshy region along the Rio Grande is upward doming of an area opposite the Rio Salado rest stop. There, surveys of terrace surfaces show that they are gradually arching upward at a rate of about 3/16 inch per year. The arching also bows up the profile of the Rio Grande. It is thought to be caused by slow upward movement of a large mass of magma below the surface. The dome-shaped uplift centers about 25 miles north of Socorro.

The Lemitar Mountains southwest of the Rio Salado are interesting in that they consist of a whole series of domino-like slices of Precambrian and Paleozoic rocks, in a pattern called listric faulting, in which each fault surface becomes less steep with increasing depth. Near the south end of the range, Paleozoic sedimentary rocks are partly covered with Tertiary lava flows and volcanic ash—the ash soft, the lava more resistant.

Elephant Head Butte, south of pointed Polvadera Peak, is the conduit of a long-extinct volcano.

Mountains on both sides of the Rio Grande gradually converge, almost closing the southern end of the Albuquerque-Belen Basin. Rocks of the Los Pinos Mountains east of the river are nearly all sedimentary—Pennsylvanian and Permian limestone, sandstone, and shale layers that show up as wavy stripes on the flanks of the range. There are a few exposures of reddish Precambrian granite at the base of the mountain escarpment, but no lower Paleozoic rocks at all. It is quite likely that they were never deposited here, that Cambrian to Mississippian seas, advancing from the west, never covered this area.

Near Socorro, the Socorro Mountains cover the western faults of the Rio Grande Rift. They consist of part of a former caldera, a collapsed volcano, later buried by other volcanic rocks, then intensively faulted and eroded so that the original caldera shape is completely gone. The whitish band part way up the mountain front marks some playa deposits formed within the rift valley, now high above it because the floor of the rift has dropped since they were deposited. Socorro Peak—the one with the "M"—is a dome of thick, viscous lava.

Socorro grew up as a mining and smelting center. Silver was mined on Socorro Mountain nearby, but most of the ores processed here came from mines farther into the mountains. Perlite, volcanic glass with little round beady structures, is mined southwest of town. After crushing and heating, it pops like popcorn, and is then used as lightweight aggregate.

Appropriately, the town is the site of the New Mexico Institute of Mining and Technology—hence the M on Socorro Peak. The New Mexico Bureau of Mines and Mineral Resources, headquartered on the campus, is a good source for further information about New Mexico geology. The mineral museum is open to the public.

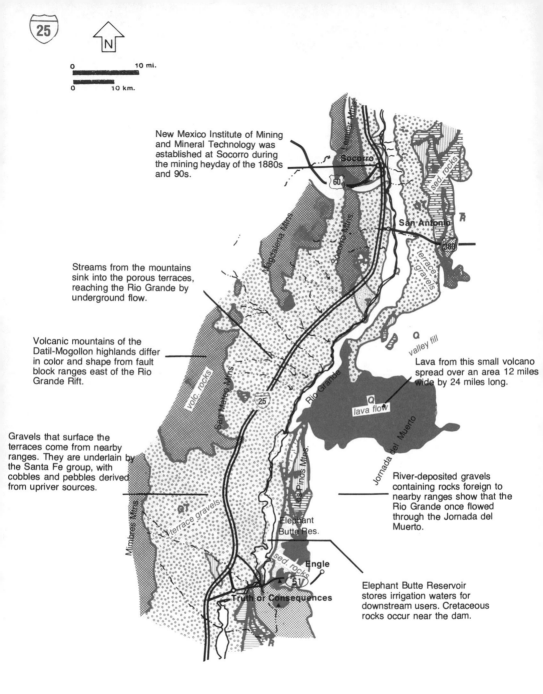

New Mexico Institute of Mining and Mineral Technology was established at Socorro during the mining heyday of the 1880s and 90s.

Streams from the mountains sink into the porous terraces, reaching the Rio Grande by underground flow.

Volcanic mountains of the Datil-Mogollon highlands differ in color and shape from fault block ranges east of the Rio Grande Rift.

Gravels that surface the terraces come from nearby ranges. They are underlain by the Santa Fe group, with cobbles and pebbles derived from upriver sources.

Lava from this small volcano spread over an area 12 miles wide by 24 miles long.

River-deposited gravels containing rocks foreign to nearby ranges show that the Rio Grande once flowed through the Jornada del Muerto.

Elephant Butte Reservoir stores irrigation waters for downstream users. Cretaceous rocks occur near the dam.

# Interstate 25
## SOCORRO—TRUTH OR CONSEQUENCES

# I-25
# Socorro—Truth or Consequences
### 74 mi./119 km.

From Socorro southward the faults along the west edge of the Rio Grande Rift are staggered, so successive mountain ranges come into view: first the south end of the Socorro Mountains, then the Magdalena, San Mateo, and Mimbres mountains. The interstate remains west of the river on terraces and alluvial fans that border the Rio Grande Valley. Both terraces and fans are veneered with gravel from nearby ranges. Arroyos and highway cuts let us glimpse the rocks beneath the veneer. Alluvial fans are made up of coarse sand and rock fragments brought in by local mountain streams, and therefore they contain only rock material derived from nearby mountains. Terraces contain sand and gravel brought into the area by the Rio Grande, with rock types from sources to the north. Some terrace surfaces are pediments, erosional features that consist of solid bedrock of mountain blocks beveled off as mountain margins are whittled back by erosion.

Distinction between alluvial fans, terraces, and pediments sounds simple, but there are complications. Some terraces in this area contain stream gravels, river gravels, *and* beveled bedrock. In places the surface gravels are so thick that one can't easily determine what underlies them. Continued faulting has raised some of the older fans and terrace deposits, and erosion has pared them off, creating new pediments that are not really related to adjacent mountains.

Nevertheless, the different surfaces here have been carefully studied and defined and named according to their elevation above the Rio Grande. Thus, the Cañada Mariana surface is about 50 feet above the river, the Valle de Parida surface is 150 feet above the river, and the Tio Bartolo surface is about 250 feet above the river. Of these, the Cañada Mariana surface is a true terrace deposit—an old Rio Grande floodplain—whereas the others are in most places erosion surfaces or pediments. Terraces also border large tributaries of the Rio Grande, and developed apparently in response to the Rio Grande's history as reflected in its own terraces.

Mountains visible across the Rio Grande Valley are fault block ranges consisting primarily of tilted Pennsylvanian and Permian sedimentary rocks; in some of them Precambrian igneous and metamorphic rocks are exposed also. Beyond these ranges is the Jornada del Muerto, a broad valley that is considered part of the Rio Grande Rift. Despite its lack of water, in Spanish days it was the route of choice from Rincon to San Antonio—shorter and safer from Apache attack. Valley deposits containing rock fragments from northern New Mexico tell us that much earlier the young Rio Grande took the same shortcut.

Black Mesa, in view across the Rio Grande from the rest stop at milepost 110, is topped with a single three-million-year-old basalt lava flow. Obviously quite fluid when it flowed from its vent, the little peak near the center of the flow, the lava is much more resistant than terrace deposits on which it lies, and protected an older terrace surface that elsewhere has been eroded away.

West of the Rio Grande, most of the rocks are volcanic. We've already met with some of them in the Socorro Mountains; others occur in the Magdalena and San Mateo mountains to the southwest, parts of the vast Datil-Mogollon volcanic field that extends westward to Arizona. Volcanoes in this area began to erupt 24-26 million years ago. Their lavas were quite different from those of Black Mesa, lighter in color (gray or purplish gray) and containing a higher percentage of silica. Most of the volcanic rocks in this region are intermediate in nature between silicic and basaltic.

Since silicic lava is thick and sticky and doesn't flow as readily as molten basalt, individual lava flows are short and stubby. Commonly they alternate with layers of tuff and volcanic breccia, products of explosive eruptions. Really cataclysmic explosions may expel so much froth, gas, and rock material that the volcanoes collapse into the partly emptied magma chambers underneath the mountains. Several calderas created by such collapses have been recognized in this volcanic field, but none are as beautifully displayed as the much younger Jemez caldera in the Jemez Mountains farther north.

South of milepost 105, you'll occasionally glimpse Elephant Butte Reservoir, part of the Bureau of Reclamation's Rio Grande Project. Wetlands near and north of this reservoir are famous for their abundant and varied birds. Mountains visible to the east, across the Rio Grande Valley, show in their coloring and weathering characteristics the contrast between the igneous intrusive rock and sedimentary rock of which they are made.

For an interesting side trip leave the interstate near milepost 83, cross Elephant Butte Dam, and follow the river road down to Truth or Consequences. A few miles east of the interstate, the community nestles among gray travertine hills, product of thermal springs that gave it its original name: Hot Springs.

*Fault block ranges along the Rio Grande wear the stripes that indicate sedimentary rocks.*

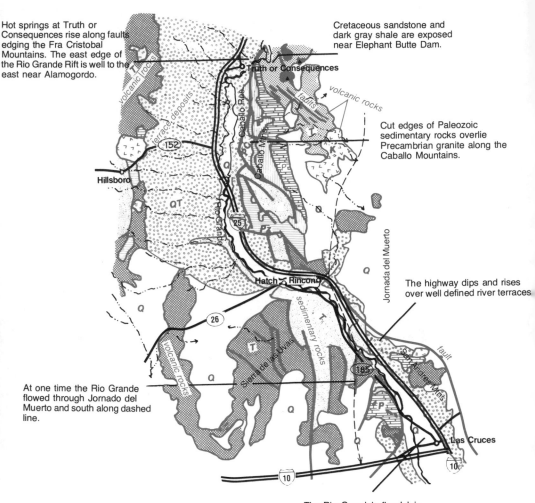

Hot springs at Truth or Consequences rise along faults edging the Fra Cristobal Mountains. The east edge of the Rio Grande Rift is well to the east near Alamogordo.

Cretaceous sandstone and dark gray shale are exposed near Elephant Butte Dam.

Cut edges of Paleozoic sedimentary rocks overlie Precambrian granite along the Caballo Mountains.

The highway dips and rises over well defined river terraces

At one time the Rio Grande flowed through Jornado del Muerto and south along dashed line.

The Rio Grande's floodplain widens near Las Cruces. Flood control embankments below the Organ Mountains protect the city from severe runoff.

**Interstate 25**
**TRUTH OR CONSEQUENCES—LAS CRUCES**

110

# I-25
# Truth or Consequences—Las Cruces
## 80 mi./129 km.

Before leaving the Truth or Consequences area you may want to explore Elephant Butte Lake and Dam. The dam stores water for irrigation in fields and farms of the southern Rio Grande Valley, and also furnishes electricity for this area. It is named for the elephant-shaped butte that juts from the lake just above the dam—a volcanic neck that can be approached by crossing the dam.

Note the columnar jointing of the rocks that make up the butte. Such joints develop at right angles to cooling surfaces, and, as you can see, the cooling surfaces must have been quite irregular. Rocks on the east bank are Cretaceous sandstone, black shale, yellowish siltstone, and limestone.

Hot spring deposits surround the town of Truth or Consequences, where water heated by contact with hot rock escapes to the surface along small faults. Because river valleys lessen the distance between sources of heat and the surface, hot springs are more likely to occur along valley bottoms, as they do here. The young volcanic features along the rift valley suggest that plenty of heat exists at shallow depths.

South of Truth or Consequences, the west edge of the Rio Grande Rift lies along the base of the Animas Range. This is the wide part of the rift, and its eastern edge is 60 miles from here, in the Sacramento Mountains east of Alamogordo. The faults that edge the west side of the rift are staggered in such a way that sloping ramps lead down into the rift valley or up out of it. Faulting of this nature suggests that in addition to simple rift-faulting due to east-west tension in the crust, there was a certain amount of counterclockwise twisting or wrenching, with the crust east of the rift moving northward relative to that on the west side of the rift.

*The Caballo Mountains, seen here across Caballo Reservoir, show southeast-dipping Paleozoic strata. The foothills are old Tertiary Santa Fe Group terraces now dissected by numerous streams.*

Sedimentary rocks in general dip east on both sides of the Rio Grande Valley. The tilt shows up particularly well on Animas Peak west of the highway. In the rugged up-faulted west face of the Caballo Mountains, east of Caballo Reservoir, contorted Paleozoic sedimentary rocks overlie Precambrian granite.

The highway skirts the inner valley of the Rio Grande as far as the south end of Caballo Reservoir, dipping and rising over an irregular surface of coarse, easily eroded gravels of the Santa Fe formation. Highway cuts show the make-up of these sediments, some of them derived from rocks far to the north up the Rio Grande.

South of the reservoir the highway drops down toward the floodplain, then swings eastward with the river to go around Tertiary volcanic rocks of the Sierra de las Uvas. Roadcuts near mileposts 52 and 53 show yellowish and reddish Paleozoic, mostly Pennsylvanian, strata. The valley is lopsided here, with big alluvial fans and steplike terraces to the west, and the Caballo Mountains rising steeply to the east.

South of Hatch the highway roller-coasters across terraces that border the valley. There is a relatively broad gap between the Caballo Mountains to the north and the San Andres Mountains to the southeast, a gap that is the southern end of the Jornada del Muerto. This wide, high valley, part of the Rio Grande Rift, was once the course of the Rio Grande. During late Tertiary time, the river flowed almost straight south through the Jornada, crossed its present course south of Rincon, and continued straight south toward the Mexican border. As fault movement deepened its present valley, the river left the Jornada del Muerto. Farther south, volcanic activity helped to deflect it into the deeper route. Today the Jornada is a closed valley, with no through drainage to the Rio Grande.

From Hatch or Rincon NM 185 forms an attractive alternate route to Las Cruces, following the west bank of the Rio Grande. For much of the route that highway is on the Rio Grande floodplain; remnants of Tertiary alluvial fans can be seen high up to the west. Tertiary volcanic rocks narrow the channel about halfway to Las Cruces.

The San Andres Mountains east of the Jornada del Muerto are composed of Precambrian igneous and metamorphic rocks overlain by Paleozoic sedimentary rocks lifted along faults. But even they are not the edge of the rift, which lies still farther east across the Tularosa Valley.

Rincon, whose name means "corner" or "angle" in Spanish, lies where the Rio Grande angles southward once more. From the top of the terraces near Rincon we get a good view northwest up the river and north up the Jornada del Muerto. There is lots of wind erosion here, where southwest spring winds whip across the Rio Grande Valley and up the Jornada del Muerto, blowing away sand and silt from natural surfaces and the plowed fields of the Rio Grande floodplain.

South of milepost 19, the mountains across the Rio Grande consist of west-dipping Paleozoic sedimentary rocks. Below them are stream-deposited and then stream-dissected fans heading in these mountains. The floodplain of the Rio Grande widens south of milepost 15, and supports farms and orchards.

Near Las Cruces a high manmade embankment just east of the highway protects the city from storm runoff from the Organ Mountains, whose "organ pipes" are made of vertically jointed Tertiary granite lifted along a major fault. Las Cruces lies on the floodplain and lower terraces that slope broadly from the Organ Mountains. The climate here in southern New Mexico is semiarid, with hot, dry summers, broken by occasional thunderstorms, alternating with cool, dry winters. Total annual rainfall is less than 10 inches. When the Spanish came here, the region was grass-covered. As cattle and sheep grazed away the grasses, desert shrubs like creosote bush, mesquite, and cacti became increasingly common. Now, even if grazing were abruptly ended, the soil could no longer support lush growths of grass. So man caused much of the desertification here.

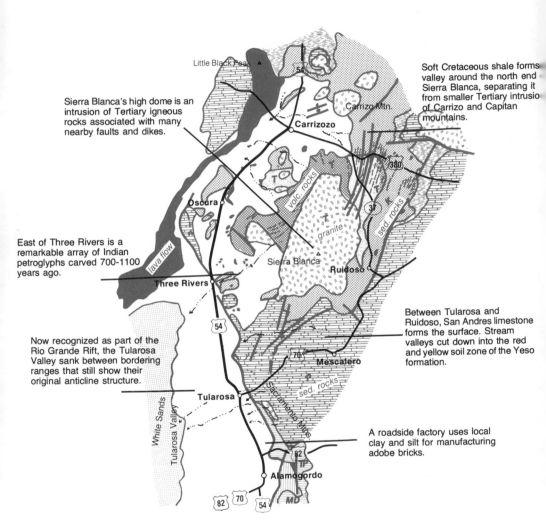

Sierra Blanca's high dome is an intrusion of Tertiary igneous rocks associated with many nearby faults and dikes.

Soft Cretaceous shale forms valley around the north end Sierra Blanca, separating it from smaller Tertiary intrusio of Carrizo and Capitan mountains.

East of Three Rivers is a remarkable array of Indian petroglyphs carved 700-1100 years ago.

Between Tularosa and Ruidoso, San Andres limestone forms the surface. Stream valleys cut down into the red and yellow soil zone of the Yeso formation.

Now recognized as part of the Rio Grande Rift, the Tularosa Valley sank between bordering ranges that still show their original anticline structure.

A roadside factory uses local clay and silt for manufacturing adobe bricks.

# US 54
# ALAMOGORDO—CARRIZOZO
## US 70   NM 37
## TULAROSA—US 380

114

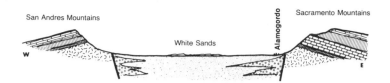

*Section across Tularosa Valley near Alamogordo.*

# US 54
# Alamogordo—Carrizozo
### 58 mi./93 km.

Driving north along this highway one can appreciate the real expanse of the White Sands. Less than half of the dune area is included within White Sands National Monument. If the day is windy, with swirls of gypsum dust rising from dunes, you can taste the sweet flavor of the gypsum!

To the west beyond the dunes are the San Andres Mountains, marked with stripes of Paleozoic sedimentary rock. The northern Sacramento Mountains to the east wear these stripes as well, as do foothills nearer the highway. Tularosa Valley, created by tension or pull-apart faulting, is part of the Rio Grande Rift. Faults that edge it are about 30 million years old, as are those of the main part of the rift, and the oldest of the sediments that fill it are of about the same age.

The valley started out as a large anticline, with Paleozoic sedimentary rocks arching across it from the present San Andres Mountains to the present Sacramento Mountains. Then as the crust pulled apart, the center of the arch collapsed, forming the valley and leaving the two ranges, the flanks of the original anticline, with faulted edges of sedimentary rock exposed. The Tularosa Valley has no outlet; water entering it from surrounding mountains sinks into the porous gravel of its floor or pools in a low area southwest of the White Sands.

The high mountain to the north, east of the highway, is Sierra Blanca, a large Tertiary intrusion surrounded with volcanic rocks.

Five miles east of Three Rivers, on county road 579, there is a remarkable mile-long array of prehistoric rock art—pictures and

*Indian petroglyphs pecked with stone tools may date back as much as 1000 years.*

geometric designs pecked into the desert varnish on volcanic rock on a lava ridge. The designs are similar to those found in pottery of the Mimbres culture of southwestern New Mexico, dated between 900 and 1300 A.D.

From Three Rivers northward a rough, black lava flow occupies the center of the valley—the Valley of Fire or Little Black Peak lava flow, best seen from Valley of Fire State Park on US 380 west of Carrizozo. Less than 1000 years ago the lava flow erupted from Little Black Peak, a tiny volcanic center near its north end. Lava flows often follow stream valleys; later the flows tend to become ridges as softer rock bordering them is worn away. New Mexico has many examples of such reversed topography; here is one in the making, though with the present climate and with surface runoff as rare as it is here, the "making" is going to be pretty slow.

Hidden beneath the lava and other valley fill are the faults that edge the east side of the rift. The fault zone curves eastward here, passing through the saddle between the San Andres and Oscura mountains.

Desert washes coming from the Sierra Blanca highlands convey storm runoff from mountains to valley. Little of their water ever reaches the low part of the valley, however: Most sinks into the broad alluvial aprons that surround the mountains.

Near Oscura we come into mining country. Gold is found in these hills—not as much as farther north near White Oaks, Nogal, and Lincoln, but enough to be worked. The earliest strikes were placer deposits—stream gravels that contained flecks and nuggets of gold and that could be worked by washing the gravels over riffle boards designed to catch the heavy bits of gold. Water of course was a problem here—it had to be carried in barrels from nearby mountains. Later mining involved more digging, but still used water to concentrate the gold. Small mine dumps near Oscura are visible from the highway.

Carrizozo with its railroad was the supply and shipping center for mines near Oscura, White Oaks, and the Nogal-Lincoln area to the east.

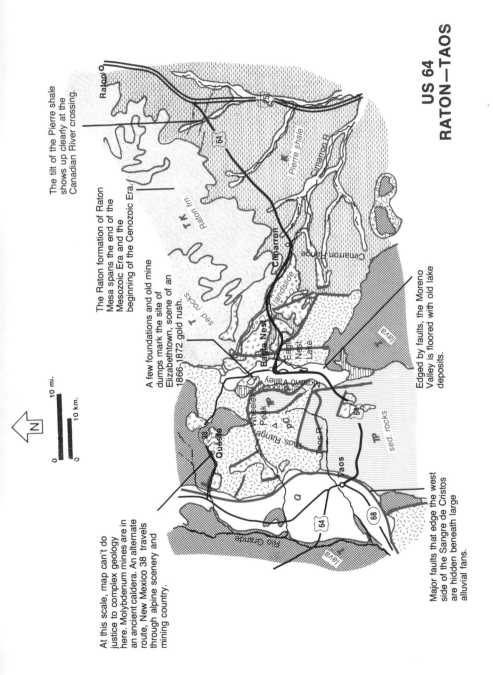

## US 64
## RATON—TAOS

The tilt of the Pierre shale shows up clearly at the Canadian River crossing.

The Raton formation of Raton Mesa spans the end of the Mesozoic Era and the beginning of the Cenozoic Era.

A few foundations and old mine dumps mark the site of Elizabethtown, scene of an 1866-1872 gold rush.

At this scale, map can't do justice to complex geology here. Molybdenum mines are in an ancient caldera. An alternate route, New Mexico 38 travels through alpine scenery and mining country.

Edged by faults, the Moreno Valley is floored with old lake deposits.

Major faults that edge the west side of the Sangre de Cristos are hidden beneath large alluvial fans.

Raton

Pierre shale

Cimarron R

TK

Raton fm.

sed. rocks

Cimarron

Cimarron Range

landslide

Eagle Nest

Eagle Nest Lake

Moreno Valley

lava

Wheeler Peak

TP

TP

Taos Range

sed. rocks

Taos R.

Questa

68

TP

Taos

Q

64

68

Rio Grande

lava

N

10 mi.

10 km.

0

0

*Dark gray sills penetrate between layers of sedimentary rock near Raton.*
W.T. Lee photo courtesy of U.S. Geological Survey.

# US 64
# Raton—Taos
### 89 mi./143 km.

Leaving Interstate 25 about eight miles south of Raton, US 64 crosses the western part of the Raton Basin before tackling the Sangre de Cristo Mountains, here divided into the Cimarron and Taos ranges. The highway at first follows the mountain branch of the old Santa Fe Trail.

The broad valley south of Raton is floored with Cretaceous shale— the Pierre formation. This soft, dark gray marine shale, which in places contains beautifully preserved fossil shells, can be seen in gullies just after leaving Interstate 25.

Between Raton and Cimarron the highway crosses three flat-topped, terrace-like pediments, erosion surfaces cut into slightly tilted Pierre shale. Where US 64 crosses the Canadian River, the tilt of this formation and its beveling by erosion show up clearly.

Small, isolated buildings along this route are pumping stations that move water from Eagle Nest Lake, in the Sangre de Cristos, to Raton. Volcanic highlands visible to the east are more than matched by steep, high bluffs flanking Raton Mesa to the west. The bluffs are capped with resistant sandstone of the Raton formation—rock deposited just at the end of Cretaceous time. Small landslide scars reveal dark gray and brownish gray shales that form parts of this unit. Deposited near shore and in lagoons and estuaries, the formation coarsens westward, toward the source of its sediments—the rising mountains. Here, shales predominate over sandstones; farther west, sandstone layers increase in number and thickness.

Bluffs here illustrate well an erosional pattern quite common in New Mexico and other arid regions: differential erosion of hard and soft rock layers. Resistant sandstones form cliffs and ledges; less resistant shale, siltstone, and coal form slopes.

There are a number of abandoned coal mines in the Raton formation and overlying units; most of them are out of sight up canyons cut in the mesa flanks. Coal mined here was used by local ranchers and the railroad or shipped south to Fort Union and Santa Fe. The mines closed after World War II.

Eagle Tail Mountain, the dark peak south of milepost 285, is made of Pliocene lava flows capped with cinder cones. The lava flowed down a deep valley, displacing the stream. Later, ridges that bordered the valley eroded away, leaving the lava as a higher topographic feature—another expression of differential erosion.

As the route drops into the valley of the Vermejo River, crossing it at milepost 270, the Cimarron Range is in view ahead. Its summits reach 11,000 to 12,500 feet. This range is part of the greater range of the Sangre de Cristos, which in turn are the southernmost end of the Rocky Mountains. The Santa Fe Trail forded Vermejo Creek about 0.8 mile west of the US 64 bridge; here we leave it to head up the Cimarron River into the mountains.

West of the town of Cimarron, as the highway curves northward at milepost 253, the Raton formation appears in cliffs to the right. Here the formation contains more cliff-forming sandstone and less slope-forming shale than we saw farther east. Along with westward coarsening, it also cuts more and more deeply into formations below, suggesting that the mountains had begun to rise and erode at the time it was deposited. The lowest Raton unit, a coarse conglomerate, reflects the very beginning of mountain uplift near the end of Cretaceous time.

On up the Cimarron is more dark gray Pierre shale. Shale of the Pierre formation is weak and erodes easily, causing many slumps and small landslides. There is one such slide area near mileposts 249 and 250, with crumpled strata near milepost 249. South of the river the entire slope is landslide material, but there it is difficult to see because of the trees.

The valley of the Cimarron River opens out at Ute Park, which is floored with silty soil derived from the Pierre shale. Baldy Peak, well north of the highway up the valley of Ute Creek, is 12,441 feet in elevation. The famous Aztec Gold Mine lay in its glacier-cut valley just about at timberline. Gold was mined from conglomerate at the base of the Raton formation, where it had settled into crevices between pebbles and cobbles, just as it does in more recent placer gravels.

Visible on the mountain ridge between Baldy and the highway is a mass of broken rock that moves slowly downhill, lubricated by ice between the rocks—a rock glacier.

Just west of the community of Ute Park are three sharp ridges, hogbacks. One is steeply tilted, resistant Dakota sandstone, the lowest Cretaceous formation here. Two others are equally resistant sills—sheets of igneous rock intruded between sedimentary layers. The light gray rock of the sills, speckled with white feldspar and black amphibole, can be seen also at the Cimarron Palisades west of milepost 243. The Palisades are cut by vertical cooling joints in the sill, which is part of an intrusion that extends southeast from Baldy Peak to the eastern edge of the mountains.

Upstream from the Palisades, watch for some very dark but glittering rock below the sill. This coarsely crystalline rock is gabbro, the intrusive equivalent of basalt. Only rarely seen at the Earth's surface, it is derived directly from the Earth's mantle. The gabbro is Precambrian, as are rocks around it: ancient volcanic rocks metamorphosed into greenstone and schist, possibly representing an ancient mid-ocean ridge. Two miles farther west the road comes into lighter colored granite and quartzite, also part of the mountain core and also Precambrian in age.

At the mouths of several tributary streams the highway cuts through coarse, cobbly gravel of several old alluvial fans. Rounded cobbles and pebbles are good evidence that the gravel is stream-deposited rather than landslide deposited. Bouncing along in a stream inevitably rounds the corners of rock fragments.

Just below Eagle Nest Dam (milepost 237), the highway veers to the right, climbing through a small patch of Cretaceous sedimentary rock. From milepost 236 we look down on Eagle Nest Lake in the Moreno Valley, a fault-edged valley with its eastern side dropped more than its western side. With ample sources of sand and gravel in the mountains around it, the valley is partly filled with Quaternary sediments. Due west is Wheeler Peak, highest point in New Mexico, part of the Taos Range. This peak and summits north of it are composed of Precambrian granite and metamorphic rocks, some of them thrust eastward over younger rocks. South of here the range was not lifted as high, and Pennsylvanian sedimentary rocks still cover the Precambrian granite.

Eagle Nest Lake provides water for Raton and nearby farms and ranches. When the reservoir is full, the lake surface is 8218 feet above sea level. Note the broad alluvial fans that reach into the Moreno Valley from surrounding highlands.

The upper Moreno Valley, north of the community of Eagle Nest, saw boom days back in 1867-1872, following discovery of gold in 1866. Elizabethtown, about four miles north on NM 38, no longer exists. A few crumbling walls and foundations mark the original townsite, while piles of bouldery gravel show the old diggings. Dumps near Elizabethtown contain galena and pyrite crystals as well as molybdenite, rhodochrosite, fluorite, and other minerals. Good mineral hunting. For those interested in mining, minerals, and alpine scenery, NM 38 continues over Red River Pass and through Red River and Molybdenum, near some modern open-pit molybdenum mines.

US 64 curves southward for some miles, crossing alluvial fans and valley fill, and then turns west again across Palo Flechado Pass (9101 feet) and the southern part of the Taos Range, separated from the higher northern part of the range by several east-west cross faults. Pennsylvanian limestones and sandstones cover the low southern end of the range. As the route drops into the canyon of Rio Fernando de Taos, outcrops reveal these rocks—dark siltstone and shale deposited in a shallow Pennsylvanian sea. Patches of Precambrian and Tertiary intrusive rocks are visible in places.

As we descend the western, wet side of the range, luxuriant forests conceal much of the geology. Notice the difference in vegetation on north and south slopes—a difference with geologic implications, in that vegetation to some degree controls erosion.

Emerging from the mountains at Taos, the highway crosses the very large faults that edge the Rio Grande Rift. Like its smaller partner, the Moreno Valley, the Rio Grande Rift is partly filled with sediments derived from mountains that border it. However the faults that edge it are much deeper, reaching down through the Earth's crust to the mantle. Mantle-derived basalt rose along the faults to erupt as volcanoes and lava flows, some of which surface the Taos Plateau. Across the valley are the Tusas Mountains, made up of Precambrian granite lifted high along the faults that edge the rift on the west.

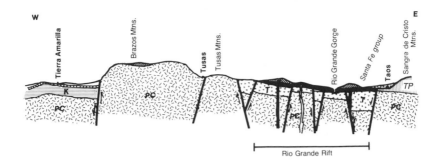

W

Tierra Amarilla

Brazos Mtns.

Tusas
Tusas Mtns.

Rio Grande Gorge

Santa Fe group

Taos

Sangre de Cristo Mtns.

E

K

PC

PC

T

T

PC

TP

PC

Rio Grande Rift

*Section along US 64 Taos – Chama*

# US 64
# Taos—Chama
### 91 mi./146 km.

Taos lies on the floodplain of the Rio Pueblo de Taos and Rio Fernando de Taos, below the west edge of the Taos Mountains. North of town, US 64 crosses wide sage-covered alluvial fans at the base of the mountains. The area near Taos is the most rapidly subsiding part of the Rio Grande Rift, and these fans are creased by low fault scarps only a few thousand years old. The rift splits the southern end of the Rockies into two prongs, the Sangre de Cristo Range, which includes the Taos Mountains, and the Tusas and Brazos mountains to the west. Total relief between the summits of the Sangre de Cristos and the real floor of the rift valley is three to five miles.

US 64 turns northwest just a few miles north of Taos, crossing lower parts of the alluvial fans. Cerro de las Ollas, a rounded mountain to the north, and a number of other domes and cones are parts of the Taos Plateau volcanic field.

Northwest of milepost 194 we begin to see a dark furrow in the light-colored sagebrush flats—the gorge of the Rio Grande. The river has been pushed westward by the growth of the alluvial fans on which the highway is built: Its gorge lies right at their western edge.

A stop at the high bridge, where the river has cut down through layered lava flows of the Taos Plateau, is well worthwhile. The lava layers exposed in the walls of the 650-foot gorge are broken by columnar joints that formed as the lava cooled. Reddish baked soil zones show between some layers.

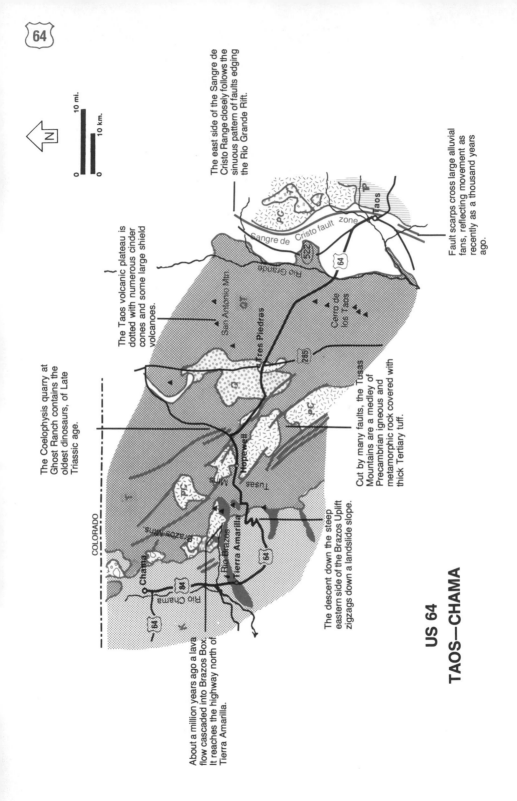

The east side of the Sangre de Cristo Range closely follows the sinuous pattern of faults edging the Rio Grande Rift.

Fault scarps cross large alluvial fans, reflecting movement as recently as a thousand years ago.

The Taos volcanic plateau is dotted with numerous cinder cones and some large shield volcanoes.

The Coelophysis quarry at Ghost Ranch contains the oldest dinosaurs, of Late Triassic age.

Cut by many faults, the Tusas Mountains are a medley of Precambrian igneous and metamorphic rock covered with thick Tertiary tuff.

The descent down the steep eastern side of the Brazos Uplift zigzags down a landslide slope.

About a million years ago a lava flow cascaded into Brazos Box. It reaches the highway north of Tierra Amarilla.

## US 64
## TAOS—CHAMA

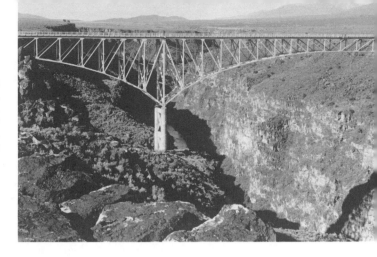

*Highway US 64 crosses the Rio Grande where it has carved a vertical-walled chasm through the superimposed lava flows of the Taos Plateau.*

Leaving the bridge, the highway climbs through roadcuts in blocky, caliche-whitened basalt dotted with vesicles or bubble holes. Small volcanoes near the road are the sources of some of these flows. There are literally hundreds of volcanoes in this volcanic field; only a few are shown on the accompanying map.

Views of the Taos Range show the deep canyons of the Rio Pueblo de Taos, the Rio Hondo, and the Red River. Mine dumps visible up the Red River's canyon mark big molybdenum mines. Note the large rounded outcrops of Precambrian granite near Taos. From this distance one can clearly distinguish the difference between the high Precambrian northern part of the range and the lower southern part, south of Taos, where Pennsylvanian sedimentary rocks form the surface.

The highway goes north around the Cerros de los Taos, two overlapping shield volcanoes composed of andesite, a somewhat stiffer lava than the surrounding basalt. The andesite contains little crystals of olivine, unusual in this kind of rock.

Close to Tres Piedras the highway crosses the faults that edge the rift valley. Tres Piedras got its name from big granite boulders that formed in place by weathering along intersecting joints. Farther west, there is more of this granite, all of it weathering the same way, in true granite fashion. The rock's rosy color comes from feldspar minerals which have turned reddish due to bombardment by radioac-

*A hardened crust to some extent protects residual granite boulders near Tres Piedras. Once they have penetrated the crust, wind and rain carve hollows in the boulders.*

125

tive decay particles. Some of the granite is case-hardened by mineral substances leached to its surface by evaporation of moisture. Such rock erodes into especially odd shapes where the hardened surface has been broken through. Notice the loose sand around the boulder piles—sand created by disintegration of the granite itself.

At Tres Piedras, US 64 crosses US 285 and heads into the Tusas Mountains. These mountains are cored with Precambrian granite and metamorphosed sedimentary rock, particularly quartzite. But conglomerate and sandstone rich in volcanic ash form much of the mountain surface. The volcanic ash came from eruptions in the San Juan Montains to the north and northwest. The soft, pinkish, tuff-rich sandstone appears in many roadcuts.

At the summit of the Tusas Mountains the view to the west takes in the relatively level top of the Brazos Mountains. There is more Precambrian rock west of the summit—both granite and metamorphosed sedimentary rock. Precambrian quartzite makes up Kiawa Mountain.

In the upper valley of the Tusas River, near the junction with NM 111, US 64 runs right along a fault zone that separates Precambrian metamorphic rocks from much younger granite, tuff, and conglomerate. The fault zone is fairly wide—as much as three miles. Displacement totals about 700 feet, enough to make the valley lopsided.

High rolling land atop the Tusas and Brazos mountains, such as that in the vicinity of Hopewell Lake, bears the imprint of Pleistocene glaciation. In Pleistocene time these rolling uplands were covered with an icecap that reached long frozen fingers down steep mountain canyons to elevations of about 8000 feet.

From the Brazos Overlook near milepost 144 there are good views of the summit plain and of the dramatic 2000-foot-deep Brazos Box, the cliff-walled canyon of the Rio Brazos, three times as deep as the Rio Grande Gorge near Taos! Age is written on the brow of the ancient quartzite that makes up the cliffs of this spectacular canyon. The north wall of the canyon, lifted along the Brazos fault, is higher than the overlook. A broad, grass-covered slope between the overlook and the highest cliff is a dipslope of east-dipping Tertiary strata that truncate west-dipping Mesozoic rocks.

Mesozoic rocks lie right on Precambrian rocks here with no intervening Paleozoic sediments. If there were Paleozoic layers here, they were eroded away during late Pennsylvanian and Permian time, when this area was part of the ancestral Rocky Mountains.

Several small cinder cones visible to the northeast are sources for lava flows that plunged down Brazos Canyon and out into the Chama Basin to the west, about 250,000 years ago.

*Precambrian sedimentary rock of the Brazos uplift, deeply gouged by Brazos Canyon, looks westward over the Chama Basin.*

To the west the view takes in the Chama Basin, and beyond it the Archuleta Arch, an anticline surfaced with Cretaceous rocks. Behind this arch is the San Juan Basin, a great crustal sag reaching almost to the Arizona border. Far away to the south are the Jemez Mountains, a range that encloses the caldera of a volcano that exploded and collapsed about a million years ago.

*Landslides in Mancos Shale are aided and abetted by construction of highways. Here US 64 is both culprit and victim.*

Landslides play an important part in shaping the western slope of these mountains. Rainfall is relatively heavy here, and wet rocks and soil slump and slide easily. In places glacial gravels overlie Mancos shale—a particularly weak Cretaceous rock unit that is especially slippery when wet. Highway builders chose what looked like an easy way down the mountain—a ramp-like slope that is almost entirely landslides! Even though highway cuts were beveled back at a shallow angle, they altered the natural angle of slope, so they often initiate slides, many of which damage the highway itself. Watch for other evidence of slides—hummocky topography, leaning trees, ponds, springs and marshy hollows, varied vegetation, roadcuts that show little scarps above bulging surfaces, patched pavement—as the highway zigzags down the west side of the mountains. Roadcut exposures show that the Mancos shale, along with man, is the culprit.

With thinner vegetation at lower elevations, rocks become more visible. Ledges and slopes of sandstone and shale are part of the Mesaverde group, which forms tawny hillsides west of mileposts 129 and 128. These rocks dip westward into the Chama Basin and, beyond the Archuleta Arch, into the San Juan Basin.

In this part of New Mexico, pre-Columbian Indians were the first "geologists." They quarried flint from Tertiary tuff in the southern part of the Chama Basin.

As the highway turns north the Brazos uplift comes into view again, this time from below. Note the nearly horizontal surface at the top of the Precambrian rocks, a legacy of an erosion surface developed either at the end of Precambrian time, or in Mesozoic time as the ancestral Rockies were worn away.

Brazos Box shows up clearly from the junction of US 64 and US 84. As we saw from the summit, the gorge follows an east-west fault with the downdropped block to the south. There are other box canyons along the edge of the uplift, but none as large and spectacular as that carved by the Rio Brazos.

The lava flow that cascades down into Brazos Box reaches the highway just north of Tierra Amarilla; its dark, columnar basalt shows in roadcuts. The town's name refers to yellowish soil derived from weathered Mancos shale.

The highway proceeds northward on the fertile floodplain of the Rio Chama and the terrace that borders it. North of milepost 334, Chama Peak comes into view. Bluffs west of the town of Chama expose Mesaverde group sandstone and shale.

Chama is a terminal for the Cumbres-Toltec narrow-gauge railway, an old mining route which now carries summer visitors across Cumbres Pass into Colorado. The Cumbres Mountains are composed of Precambrian granite and Tertiary volcanic rocks.

**US 70/ US 82**

**LAS CRUCES—ALAMOGORDO**

Paleozoic strata stripe the face of the Sacramento Mountains, the eastern part of an anticline once continuous across the Tularosa Valley.

The Tularosa Valley is now recognized as part of the Rio Grande Rift.

US 54 passes close to the Jarilla Mountains, site of an 1880s miniature gold rush. Water for concentrating the gold was always a problem here.

Ores found along this fault led to mining of lead, silver, copper, and gold. This may be the location of the fabled Lost Padre Mine.

White Sands' wind-deposited gypsum comes from playas and gypsum-saturated soil just west of the dunes.

The ultimate source of White Sands gypsum is in Permian rocks of the Sacramento Mountains.

Dots outline old calderas thought to be present in the Organ and Dona Ana mountains.

Sacramento Mtns.

Alamogordo

Orogrande

dunes

White Sands NM

Alkali Flat

Lake Lucero

dunes

San Andres Mtns.

Tularosa Valley

valley fill

San Augustin Pass

intrusive rocks

volc. rocks

Organ Mine

Organ

Dona Ana Mtns.

Las Cruces

N

10 mi.

10 km.

130

# US 70, US 82
# Las Cruces—Alamogordo
### 66 mi./106 km.

Leaving Interstate 25 at the north edge of Las Cruces, the highway crosses a floodwater revetment that prevents storm runoff from flooding the city. The highway then climbs three poorly defined terraces—old Rio Grande floodplains—to the southern end of the Jornada del Muerto, a broad, dry valley that was the preferred route for Spanish travelers going north to Santa Fe. Even though it is separated from the main channel of the Rio Grande by small faulted and intruded ranges, the Jornada is considered part of the Rio Grande Rift.

Normally hidden by a thin veneer of gravel, sediments that make up the terraces can be seen in roadcuts and gravel pits about a mile east of Interstate 25. They are part of the Santa Fe group, found all up and down the Rio Grande Valley. In this area they consist of old alluvial fans interlayered with river sediments that tell us that the Rio Grande at one time flowed through the Jornada del Muerto, before shifting to its present channel farther west.

In places, the Santa Fe sediments contain petrified bones of fossil mammoths, mastodons, horses, camels, and other vertebrates that wandered across this area between ten million and five hundred thousand years ago. The porous gravels serve Las Cruces well, acting as aquifers for the city's water wells. A little farther north, the road crosses an almost imperceptible drainage divide, north of which both surface and subsurface water flow toward the Jornada del Muerto basin.

As the road crosses the south end of the Jornada del Muerto, there are good views of the Organ Mountains. The range gets its name from craggy, organlike outcrops of Tertiary granite, 27 million years old, that make up its steep eastern face. On the mountain the shape of the intrusion is almost perfectly outlined. Its many joints, developed as the granite cooled and shrank, were probably enhanced by movement along rift faults.

A few slivers of Paleozoic sedimentary rock edge the base of the mountains; there is a little fluorspar mine near their line of contact with the intrusion. Fluorspar is the principal source of fluorine, and is used as a flux in smelters, in preparation of glass and enamel, and in manufacturing hydrofluoric acid and fluorocarbons. Fluorspar is known to occur in association with calderas—collapsed volcanoes. Two such structures have been identified here, ringed with faults and associated with Tertiary intrusive and volcanic rocks of the Organ and Dona Ana mountains.

Old mines near Organ, along a fault where quartz veins cut through Precambrian rocks, produced copper, lead, silver, gold, and zinc.

As the highway climbs more steeply, look north into the San Andres Mountains to see their west-dipping bands of Pennsylvanian and Permian strata. In the foreground the west-dipping pattern is somewhat jumbled, but farther north it becomes quite systematic. There are 4700 feet of Paleozoic rocks here; the top 3000 feet of them are Pennsylvanian and Permian. Almost all the formations are

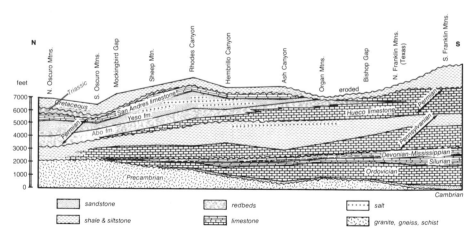

*By diagramming rocks identified in successive canyons of the Oscura, San Andres, Organ, and Franklin Mountains, geologists have clarified relationships between Paleozoic and Mesozoic formations. Wavy lines show unconformities. Pre-Permian limestones thicken southward in the direction of the open seas. Permian and Triassic strata thicken and coarsen northward, in the direction of the newly risen ancestral Rocky Mountains.* Adapted from Kottlowski.

132

marine, deposited in shallow seas that advanced over a nearly level land. Some of the Paleozoic rocks were quarried near milepost 160 for crushed stone used in the flood-control revetments in Las Cruces.

At San Augustin Pass (5719 feet) the highway cuts through the Tertiary intrusion of the Organ Mountains. Roadcuts offer a good chance for a close look at the unweathered granite. Elsewhere, the densely jointed rock is iron-stained from weathering of pyrite grains within it. The mineral pyrite, a brassy yellow iron sulfide, is sometimes called "fool's gold," so don't be fooled if you find some in this rock.

The viewpoint east of the pass looks down into the Tularosa Valley, an enclosed desert basin with no external drainage. During Pleistocene time, when rainfall was greater than it is now, the valley contained a lake known to geologists as Lake Otero. Water draining the San Andres Mountains is rich in gypsum and other salts picked up from Permian rocks, so gypsum and other salts accumulated in the lake deposits. Today the lake is gone, but a white gypsum playa called Lake Lucero remains. Strong spring winds whipping in from the southwest pick up gypsum from the floor of this playa and from the soil around it to create the dunes of White Sands National Monument.

As the highway descends into Tularosa Valley, large slide blocks of Paleozoic rock are visible to the north, their eastward dip contrasting with the predominant westward dip of rocks in the San Andres Mountains. The highway soon comes into much older rock: Precambrian granite that underlies the Paleozoic strata and that is exposed here along the fault that edges this side of the San Andres Mountains. The ancient granite is much darker than the younger granite of the Organ Mountains.

Across the valley, the Sacramento Mountains also contain Paleozoic sedimentary rocks underlain by Precambrian granite. There, the strata dip east. With the San Andres Range, the Sacramentos form an immense anticline, with the Tularosa Valley lying where its crest should be. A giant wedge running the length of the anticline sank several thousand feet to create the Tularosa Valley.

Mineral Hill, a granite hill north of milepost 167, just *may* be the location of the Lost Padre Gold Mine, worked in the late 1790s by a French priest. Its location was described as "two days north of El Paso del Norte, east of El Camino Real, in a lone mountain surrounded by a large basin, with a spring nearby."

At the base of the mountains proper, the highway crosses the alluvial apron that surrounds them. Stream gravel and sand and the finer sediments of a succession of lakes, as well as layers of salt and gypsum, fill the fault-edged valley to a depth of about 4000 feet. Soils on the floor of the valley are high in salt and gypsum, discouraging the growth of all but salt-tolerant plants.

From milepost 180 you can see the fault scarp of the mountains quite well. Low steps crossing Quaternary terraces mark faulting that obviously occurred after terrace development. Though no movement has been recorded in historic time, faults along the edges of the Tularosa Valley may well be active still.

Small buttes east of milepost 192 are lone outcrops of Paleozoic limestone, lifted along faults, jutting up through the valley fill.

In a few more miles gypsum dunes begin to appear along the roadside. Some are fairly stable and have developed growths of yucca, saltbush, and other plants. Others move too actively for vegetation to become established.

*Southeast of Alamogordo, the flanks of the Sacramento Mountains expose Paleozoic sedimentary rocks deposited on the floor of a shallow sea.*

On the west wall of the Sacramento Mountains, sliced-off edges of Paleozoic sedimentary rocks lie above Precambrian granite. The sedimentary rocks range in age from Cambrian to Permian; the highest cliff is Mississippian limestone. Geologists studying these rocks along the cut faces of the San Andres Mountains have found

134

that Cambrian, Ordovician, Silurian, Devonian, and Mississippian rock layers thicken southward, while Pennsylvanian and Permian strata thicken northward.

Variations in rock types, too, are well exposed here. The Paleozoic layers were deposited in a succession of seas that swept across a nearly flat coastal plain, a region much like the Gulf Coast today. As along modern shores, sandy beaches gave way to muddy bays, and muddy bays to gravelly river estuaries. Conglomerate and sandstone were deposited near the shore, whereas shells of marine animals accumulated offshore to become limestone. Unconformities developed when the marine sediments were raised above sea level for a time and eroded. Thus geologists can reconstruct the Paleozoic history of the area, telling when seas came and went, which direction they came from, where bays indented the shoreline or river deltas extended it, even the direction of far-off mountains. Using the diagram, try it!

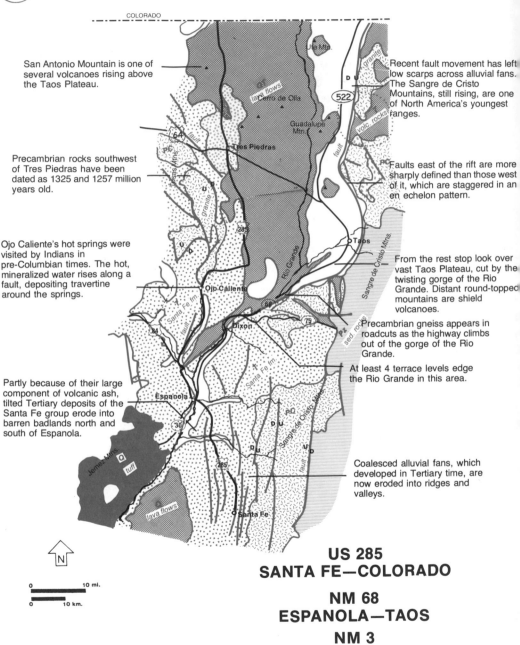

San Antonio Mountain is one of several volcanoes rising above the Taos Plateau.

Precambrian rocks southwest of Tres Piedras have been dated as 1325 and 1257 million years old.

Ojo Caliente's hot springs were visited by Indians in pre-Columbian times. The hot, mineralized water rises along a fault, depositing travertine around the springs.

Partly because of their large component of volcanic ash, tilted Tertiary deposits of the Santa Fe group erode into barren badlands north and south of Espanola.

Recent fault movement has left low scarps across alluvial fans. The Sangre de Cristo Mountains, still rising, are one of North America's youngest ranges.

Faults east of the rift are more sharply defined than those west of it, which are staggered in an en echelon pattern.

From the rest stop look over vast Taos Plateau, cut by the twisting gorge of the Rio Grande. Distant round-topped mountains are shield volcanoes.

Precambrian gneiss appears in roadcuts as the highway climbs out of the gorge of the Rio Grande.

At least 4 terrace levels edge the Rio Grande in this area.

Coalesced alluvial fans, which developed in Tertiary time, are now eroded into ridges and valleys.

**US 285**
**SANTA FE—COLORADO**

**NM 68**
**ESPANOLA—TAOS**

**NM 3**
**TAOS—COLORADO**

N

| 0 | 10 mi. |
| 0 | 10 km. |

136

# US 285
# Santa Fe—Colorado
### 106 mi./170 km.

Going north out of Santa Fe, US 285 climbs to the top of alluvial fans that make up the foothills along the west side of the Sangre de Cristo Range. These Pleistocene fans are now deeply channeled by streams from the mountains. Canyons and roadcuts reveal their fine, light orange sand interlayered in places with gravel.

Older sediments, lake and stream deposits of the Santa Fe group, appear on Camel Rock and, as we'll see, farther north. Near Camel Rock we can see that harder layers resist erosion better than softer layers—an example of differential erosion. Wind is an important factor in shaping such isolated rock remnants; bouncing tiny tools of sand up out of the Rio Grande floodplain, it attacks the lower portions of such rocks with particular vehemence.

Santa Fe group deposits are visible out in the Rio Grande Valley proper from about milepost 245 north. Tilted by rift faulting, they make up the Espanola Barrancas or Badlands. Deposited before the Rio Grande became a through-flowing river, they are loaded with volcanic ash from eruptions in volcanic centers to the north and west. Clay derived from volcanic ash contributed to badland formation because it erodes easily, swelling when wet, shrinking when dry. Plants have difficulty growing in such soils, especially in an arid climate, and wind and rain freely remove loosened clay and silt.

Silty, sandy deposits of the barrancas contain fossils of now-extinct mammals—ancestral horses, deer, camels, bears—that thrived here in the early days of the Rio Grande Rift. Roadcuts just south of Espanola show the layered ash, pinkish sandstone, and caliche—calcium carbonate deposited by groundwater—of the barrancas.

The highway drops down almost to river level at Espanola, on the fertile bottomlands of the Rio Grande. The valley is broader here than it is farther north, where resistant lava flows of the Taos volcanic field hold the river in a narrow gorge. Fields along the river are irrigated directly from the river or from tributary streams coming from the mountains.

Crossing the river north of Espanola our route follows the Rio Chama and Rio Ojo Caliente northward. The basalt-capped mesa ahead at milepost 263 is Black Mesa, the southern end of the Taos Plateau. This volcanic tableland consists of relatively thin, almost horizontal lava flows from volcanic centers strung out along the Rio Grande Rift. The flows are composed of basalt, the most fluid of lavas, which rose from the mantle through the very deep faults that border the rift. Lighter-colored hills and badlands south of the basalt are Santa Fe group sand and gravel deposits.

Black Mesa is to the north at milepost 335, with more badlands to the east. The easily-eroded Santa Fe group of the badlands is responsible for the light-colored dust and sandy soils here. Thin, hard layers in the Santa Fe group represent still-stands of the water table, with silica and caliche deposited between sand and gravel grains. In places the Santa Fe group is tilted, evidence that faulting and earth movement have continued in the Rio Grande Rift since it was deposited.

River deposits are notoriously variable, expecially with the added element of volcanic eruptions, and the Santa Fe group has been divided into a number of formations differing in rock type and position. The formations range from mid-Tertiary to early Quaternary in age.

Also in view from milepost 335, to the west, is the broad, somewhat saddle-shaped rim of the Valles Caldera, the remains of a one million-year-old volcano.

The highway turns away from the Rio Chama between mileposts 337 and 339, and rounds the northwest side of Black Mesa. Reddish rocks at the south end of Black Mesa are part of the Santa Fe group. Near milepost 340 these rocks have been eroded into mushroom rocks, figurelike hoodoos, and other unusual erosional forms; their disintegration has furnished sand for sand dunes. Watch for crossbedding in this rock. Crossbeds are short and straight, of the type characteristic of sandstone deposited by flowing water.

Black Mesa's basalt cap has been dated at 2.8 million years. Note its fine, closely-spaced columnar jointing near milepost 340. It is the last volcanic rock to flow into the rift valley before the Rio Grande was established as a through-flowing river. Some of the basalt from the Jemez Mountains area, dated at 2.4 million years, flowed across river gravels. The two dates therefore bracket the development of the through-flowing Rio Grande between 2.8 and 2.4 million years ago.

Hot mineral springs at Ojo Caliente, a mile west of milepost 353, come to the surface in a faulted area where Precambrian granite and quartzite are lifted above the Santa Fe group, defining the west side of the Rio Grande Rift. The springs, with temperatures of 98-113 degrees Fahrenheit, contain minor amounts of iron, arsenic, soda, lithium, and other minerals.

North of Ojo Caliente we rise again onto alluvial fans, with scattered outcrops of purplish sedimentary rocks derived from volcanic highlands straddling the New Mexico-Colorado border to the northwest. In the Tusas Mountains between Ojo Caliente and Petaca, about 15 miles to the north, the Precambrian granite contains bands of coarse pegmatite with large crystals of mica, beryl, garnet, feldspar, tourmaline, and other minerals. Lots of good mineral-hunting in this area.

As the highway swings eastward and onto the basalt surface of the Taos Plateau, views of the Taos Mountains show the towering Precambrian mass of the northern Taos Range to the northeast, and the lower southern part of the range, surfaced with Pennsylvanian sedimentary rocks, directly ahead. Wheeler Peak, high point in the northern part of the range, is New Mexico's highest point, too, at an elevation of 13,161 feet. The southern end of the Taos Range merges with the Truchas Mountains, whose highest peaks are also over 13,000 feet high. The Truchas Mountains end abruptly near Santa Fe—the southern end of the great Rocky Mountain chain.

The Tusas Mountains to the west are not as high, but they are both scenic and dramatic. Composed of Precambrian rock—mostly granite and quartzite—they are faulted on the east and tilted up like a trap door. The same Precambrian rocks underlie the lava flows of the Taos Plateau, but there they have dropped many thousands of feet as part of the Rio Grande Rift.

Tres Piedras is named for the picturesque granite rockpiles that surround it—geologically part of the granite of the Tusas Mountains. This rock is cut by many joints, and weathering along the joints has gradually rounded individual granite boulders—good examples of spheroidal weathering. The granite, as well as white Precambrian quartzite and reddish volcanic cinders, are used in landscaping around the ranger station.

From the hilltop north of Tres Piedras, San Antonio Peak is visible to the north. Like other high, isolated domes rising above the Taos Plateau, it is a volcano, with a breached crater at its summit. Obviously its lavas were thicker, more viscous, than the freely flowing basalts that surface the plateau.

There are many smaller cinder cones on the plateau also; some of them are being quarried for cinder or for perlite, a form of volcanic glass. When it is crushed and heated, perlite pops like popcorn; in this form it is used for lightweight aggregate and as insulating material.

Colorado's Sangre de Cristo Mountains rise to the north, behind another dome-shaped volcano, Ute Peak. North of San Antonio Mountain are a number of cinder cones quarried for their very red cinder, also used for lightweight aggregate. The Rio Grande Rift, still covered with Taos Plateau lava flows, broadens out here. Northward it becomes the San Luis Valley in Colorado.

*The ropy surface of the Valley of Fire lava flow is characteristic of very fluid basalt lava.*

# US 380
# Carrizozo—I-25
## 64 mi./103 km.

Going west from Carrizozo, the highway almost immediately comes to rugged, dark basalt lava flows known as the Valley of Fire or Little Black Peak lava flows. Little Black Peak, the small volcano from which the lava came, can be seen to the northwest. About 44 miles long, only a few miles wide, and less than 1000 years old, the lava actually represents several long, thin, overlapping flows. The surface of the lava still exhibits many of its original features: a rumpled, ropy, shiny surface formed from lava the consistency of taffy; caves or tubes created as molten lava flowed out from beneath its solidified crust; pressure ridges where thickening lava pushed up through breaks in hardened crust; domes, blisters, and tiny bubblelike vesicles formed by gases inherent in the fluid basalt; and many deep fissures resulting from shrinkage of the basalt as it cooled.

Ridges of older sedimentary rock jut through the lava flow. One of them, a hogback of Dakota sandstone, forms the Valley of Fires State Park campsite. Though the lava has not yet decomposed into soil, desert plants—for the basalt surface *is* a desert—grow in pockets of sand and silt brought in by wind. Small desert animals, many of them wearing darker-than-normal camouflage in adaptation to their dark surroundings, live among the fissured rocks and pockets of vegetation. In White Sands National Monument, only 70 miles farther south, the same animals are whiter than normal!

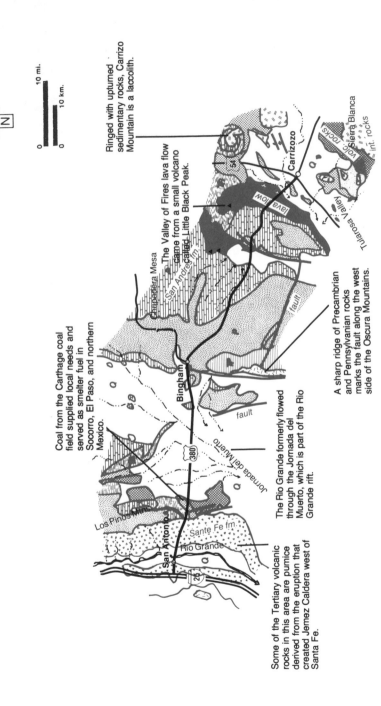

Ringed with upturned sedimentary rocks, Carrizo Mountain is a laccolith.

The Valley of Fires lava flow came from a small volcano called Little Black Peak.

Coal from the Carthage coal field supplied local needs and served as smelter fuel in Socorro, El Paso, and northern Mexico.

A sharp ridge of Precambrian and Pennsylvanian rocks marks the fault along the west side of the Oscura Mountains.

The Rio Grande formerly flowed through the Jornada del Muerto, which is part of the Rio Grande rift.

Some of the Tertiary volcanic rocks in this area are pumice derived from the eruption that created Jemez Caldera west of Santa Fe.

Chupadera Mesa

San Andres fm

lava flow

Carrizozo

VOLC. ROCKS

Sierra Blanca

int. rocks

Tularosa Valley

fault

Bingham

fault

Jornada del Muerto

380

Los Pinos Mtns

San Antonio

Sante Fe fm

Rio Grande

25

## US 380
## CARRIZOZO—INTERSTATE 25

142

*Ropy lava forms when a thin, partly cooled lava skin is moved along by more fluid lava beneath. The curve of the crinkled surface shows that this flow moved from the upper right to the lower left.*

At the western edge of the lava flow the highway climbs onto light gray Permian San Andres limestone, on which it remains, except for dry stream valleys filled with recent gravels, halfway to Bingham. This is a marine limestone, and contains many fossil shellfish. In the desert climate, groundwater saturated with calcium carbonate from the limestone is drawn toward the surface, where it evaporates, depositing its calcium carbonate in the form of caliche, the whitish, chalklike material visible in shallow roadcuts.

Red rocks of the Abo formation, also Permian, border the highway as it curves northward around the Oscura Mountains. Unlike most other Paleozoic formations in New Mexico, the Abo is continental—deposited on land rather than in the sea. It bears all the hallmarks of a delta-floodplain type of deposition. Its red color comes from oxidized iron; green seams and stringers are also tinted by iron, in this case not oxidized, possibly because of the presence of a little decomposing plant or animal material in the original sediment—in which case the decomposition bacteria use up all available oxygen.

The Oscura Mountains separate the Tularosa Valley from the Jornada del Muerto, another depression of the Rio Grande Rift sys-

*A small, dark cave leads to a lava tunnel left as fluid lava flowed out from under a cooling crust. Some lava tunnels extend for hundreds of feet.*

143

tem. Not very high as mountains go, the range is an east-tilted block of Precambrian, Pennsylvanian, and Permian rocks faulted along its western side. The fault passes right under the town of Bingham, and the sharp-prowed fault scarp can be seen to the south from just west of that community. A smaller fault block, also with east-tilted strata, is southwest of Bingham.

The course of history changed at 5:29 on the morning of July 16, 1945, when the first atomic bomb was tested in the Jornada del Muerto valley a few miles south of here—an explosion heard as far away as Flagstaff, Arizona. A lake basin in Pleistocene time (old beach ridges can be seen on satellite images), the valley near the explosion site is now a playa. The site is closed to visitors.

The Jornada del Muerto, once much deeper than it is now, is filled in with Cenozoic sediments—lake deposits, gravel washed from surrounding mountains, and some gravels unrelated to nearby mountains but brought here by the Rio Grande at a time when it flowed through this valley, before shifting to its present course. All streams that enter the Jornada del Muerto now sink into these porous gravels; the valley has no through drainage. Geophysical studies in which manmade seismic waves are recorded as they bounce off buried rocks indicate that an older range of mountains, lifted in late Cretaceous time—at the same time as the Rocky Mountains farther north—is now buried beneath these gravels.

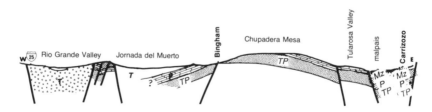

*Section along US 380 between Carrizozo and I-25*

Low ranges west of the Jornada del Muerto are three-dimensional jigsaw puzzles of Paleozoic, Mesozoic, and Cenozoic fault slices, with older rocks in places thrust over and across younger ones. Cretaceous coal was mined from this range in the 1880s and '90s and well into the 20th century, as fuel for silver and lead smelters in Socorro. More distant mountains on the other side of the Rio Grande Valley are part of the volcanic highlands that extend west into Arizona and south almost to Interstate 10.

Sand dunes near US 380, partly anchored by vegetation, derive their sand and their color from the Abo formation. Stop and pick up a handful of this sand; notice the uniform size and rounded shape of the grains, characteristic of dune sand everywhere. Dunes develop where there is plenty of sand, plenty of wind, and a surface topography over which the wind slows down and drops the sand. Here, windspeed is lessened by fault scarps of the Oscura Mountains, which change the flow pattern of winds sweeping across the Rio Grande Valley from the west.

The highway descends toward the Rio Grande on Tertiary strata made up of Rio Grande river deposits interbedded with volcanic ash from explosive eruptions near and southwest of Socorro. One recognizable ash layer stems from the eruption of the Jemez volcano west of Santa Fe, just a million years ago. This part of the Rio Grande Rift is still very active geologically, with frequent earthquakes. The rift continues to deepen as its margins continue to pull apart.

The highway descends toward the river, cutting through coarse gravel of an alluvial fan and crossing old Rio Grande deposits now trenched by the same river that created them. The Rio Grande seems to have shifted from the Jornada del Muerto to its present course about a million years ago—probably because its present valley was dropping more rapidly than other parts of the rift. Crossing the river here, the highway ascends the big alluvial fan of Nogal Canyon and joins Interstate 25 at milepost 175.

*The Rio Grande's broad waters are deceptively tame in winter. As spring warmth melts snow in the Rockies, the river becomes more turbulent despite the calming influence of upstream dams.*

145

# NM 68, NM 522
## Espanola—Colorado via Taos
### 91 mi./146 km.

For a map of this route, see US 285 Santa Fe—Colorado p. 136

For the first few miles north of Espanola this route follows US 285 along the Rio Grande floodplain and across pinkish rocks of the Santa Fe group—river sand and gravel containing lots of volcanic ash. High mesas to the west are capped with lava from the Jemez Mountains, site of a volcanic drama played out a million years ago.

Leaving US 285 north of Espanola, NM 68 follows the Rio Grande north toward Taos. Several terrace levels—old river floodplains—can now be seen on either side of the river. Badlands visible across the river are carved in the Santa Fe group.

The high plateau ahead and to the left from mileposts 6-7 is the south end of lava-capped Taos Plateau. The highway parallels its margin for some distance, crossing a number of normally dry washes draining the Truchas Range, part of the Sangre de Cristo Mountains. The plateau basalts overlie more Santa Fe group sandstone and siltstone that in many places are visibly tilted by slumping and by continuing movement along Rio Grande Rift faults.

North of Velarde the Rio Grande's valley narrows, constricted by hard lavas of the Taos Plateau. Tumbled boulders of basalt cover hills on either side; through them you can see patches and occasional pinnacles of Santa Fe group sandstone and siltstone. The basalt is many-layered and in places shows vertical cooling joints that make it look like a log stockade. Because lava flows cool from both top and bottom, shrinking at different rates, each flow breaks into two sets of stockadelike columnar joints, the upper ones usually taller and thinner than the lower.

NM 75, leaving our route at Embudo, goes to the old town of Dixon. Near there, coarse-grained pegmatite veins are mined for large crystals of beryl, columbite, mica, and other minerals.

North of milepost 23 the highway climbs part way up the side of the Rio Grande Gorge. Roadcuts reveal interesting old stream channels filled with basalt boulders. Descending again through the lava flows, the road travels below cliffs and highway cuts of Precambrian metamorphic rocks—highly jointed gneiss, schist, and slate that break into angular blocks and skid downslope toward the Rio Grande. The river flows right along the contact between these ancient metamorphic rocks and the *much* younger basalts to the west.

Near the town of Pilar we leave the Rio Grande and its narrow canyon and climb through both Santa Fe group and basalt flows to the surface of the Taos Plateau. Coarse gravel visible along the highway near milepost 29 is largely Pleistocene, washed down from the Sangre de Cristo Mountains. Notice its well rounded boulders and cobbles and the suggestion of stratification in pebble, cobble, and boulder layers.

Between mileposts 29 and 30, the uppermost basalt layer to the west is a good example of the way a single flow cools into what appears to be two layers. Immediately beneath the basalt is some old soil baked to a deep, dark red by the heat of the lava flow.

Sand eroding from Santa Fe group sandstone is being redeposited as dunes near mileposts 31 and 32. In a number of places, boulders and cobbles again fill old channels in the sandstone.

North of milepost 32 the highway emerges from the confines of the Rio Grande Gorge. From the rest stop near milepost 35, the gorge can be seen twisting across the Taos Plateau. Off to the west, not particularly prominent, is one of the volcanic centers from which the plateau lavas flowed. Extending north into Colorado, the lava covers an area nearly 100 miles long and about 20 miles wide. A number of large round-domed volcanoes rise above it north of Taos.

East of the highway on the approach to Taos, the Sangre de Cristo Mountains are lower than elsewhere. Vertical uplift is not as great there, and the mountains are surfaced with Pennsylvanian sedimentary rocks. North of Taos, where the Pennsylvanian rocks are faulted against resistant Precambrian granite and gneiss, the picture is different: Summits are much higher, and the mountain front is abrupt, forming a dramatic backdrop for the town.

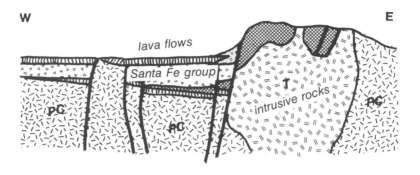

W              E

*Section across NM 3 at Questa.* (After Lipman and Reed)

Taos Pueblo is certainly worth a visit. In the shadow of their sacred mountain, Taos Indians live today very much as their ancestors lived, but with added modern amenities—cars, water, and electricity. Built of natural materials, this pueblo and others along the Rio Grande have set the style for much non-Indian architecture in New Mexico.

North of Taos the highway—now NM 522—crosses the great alluvial apron at the base of the Sangre de Cristo Mountains, paralleling the dramatic mountain front. The range rises 5600 feet above the Taos Plateau, and includes New Mexico's highest peak, Mt. Wheeler, as well as some spectacular alpine scenery. The Sangre de Cristo fault, along which the Rio Grande Rift subsided, runs between the highway and the mountains; total vertical movement here amounts to about 10,000 feet.

Stretching north well into Colorado, these mountains are one of the youngest ranges in America. Born in Miocene time, they are still rising: Low fault scarps, judged to be about 10,000 years old, cut across the alluvial apron at their base.

The mountains are composed of Precambrian metamorphic rocks about 1.7 billion years old, somewhat younger Precambrian granite, and a number of small Tertiary intrusions which seem to be finger-like projections rising from the top of a much larger, otherwise unexposed intrusion.

In the mountains east of Questa, Tertiary volcanic rocks are involved with a complex and mineral-rich caldera that formed about 25 million years ago. Calderas, circular, cliff-walled valleys, form by collapse of volcanoes into their partly emptied magma chambers. The basin shape of this caldera is long since gone, but thanks to 60 years of mining activity and unusually good exposures in the deep canyon of the Red River, its complex geology is fairly well known.

The tuff, the caldera, and the small Tertiary intrusions are all thought to be products of the large intrusion under the mountain. This large granite mass may represent the cooled and hardened contents of the magma chamber responsible for explosive eruptions that showered the area with volcanic ash and led, ultimately, to collapse of the volcano. There is evidence, furthermore, that collapse was followed by an upward surge of magma that formed a dome within the caldera itself.

Major molybdenum mines up the canyon of the Red River east of Questa obtain ores from a ring of secondary intrusions around the southern rim of the caldera.

Northwest of Questa is Guadalupe Mountain, a cluster of volcanoes rising above the lava expanse of the Taos Plateau. North of Guadalupe Mountain, a number of other volcanoes, all formed within the last 10 million years, come into view to the north and northwest. All these steep-sided volcanoes are composed of silicic lava, which is much less fluid than the basalt of the lava plateau on which they rest. Most of the volcanoes have summit craters not visible from the highway.

The route continues north to Colorado on the sloping alluvial apron at the foot of the Sangre de Cristo Mountains. Low, terracelike steps on this apron, barely visible as a change of color in the sagebrush, are fault scarps indicating fairly recent movement—within the last 10,000 years—along the fault that edges the Rio Grande Rift. Many of the mountain spurs are beveled or cut off, another sign of recent fault movement. To the north, the great mass of Blanca Peak, in Colorado, breaks the regular line of the Sangre de Cristos, which extend north to Salida, Colorado.

A circlet of lava domes and a central resurgent dome rise from the floor of the Jemez Caldera.

Many rift valley faults are not shown. These reflect north-south trend.

lava dome

A few miles north on US 84 is Ghost Ranch, site of dinosaur finds. An exhibit explains local geology.

Volcanic vents (not visible from the highway) are tidily aligned along north-south faults deep enough to tap basalt magma from the mantle.

285

Santa Fe

Pojoaque

25

Santa Fe group

Española

502

lava

Rio Grande

Hernandez

Santa Fe group

Abiquiu

lava & tuff

Los Alamos

Frijoles R.

Bandelier NM

Jemez Mtns.

Coyote

Q

sed. rocks

pC

Bandelier tuff

Jemez Springs

lava & tuff

96

fault

Pz

Q

faults

Jemez Pueblu

Gallina

sed. rocks

96

Cuba

T

pC

Pz

Pz

San Isidro

Sierra Nacimiento

Nacimiento fault

44

Cabezon Peak

sed. rocks

Weak, easily eroded Triassic and Jurassic rocks of mesa walls are capped with resistant Cretaceous sandstone.

Cabezon Peak and other volcanic necks dot the region southwest of the Jemez and Nacimiento Mountains.

N

0   10 mi.

0   10 km.

**JEMEZ MOUNTAINS LOOP**

# Jemez Mountains Loop
## 188 mi./302 km.

This loop drive is described in a clockwise direction
starting at Pojoaque, just south of Española.

Leaving US 285 near Pojoaque, NM 502 follows the Nambe River
to the Rio Grande. Views to the north show badlands etched in soft,
tan, easily eroded Tertiary sandstone, siltstone, and volcanic ash of
the Santa Fe group, tilted by movements along faults that edge the
Rio Grande Rift. To the west the rift faults are hidden by the thick
pile of volcanic rocks of the Jemez Mountains.

After crossing the Rio Grande the route climbs through more tan
Tertiary rocks to the black lava that caps them. Note the columnar
jointing of this lava. The geology is complex here, but the basic
pattern is one of Tertiary volcanic and sedimentary rocks overlain by
younger lava flows, many of which filled valleys carved in the older
rocks. In places, Pleistocene gravel fills old channels and spreads
thinly across the pre-lava surface.

Farther west, tawny pink rock becomes increasingly
prominent—fairly massive in appearance but full of small holes
eroded by wind and water. This is the Bandelier tuff, product of the
two great eruptions that shaped the Jemez Mountains of today. We'll
see more of it on our circuit of the mountains.

In the canyons below White Rock, lava flows that underlie the
entire range are well exposed. We'll see more of these rocks on our
circuit as well.

*This panorama shows less than half of sun-splotched Valles Caldera. Lava domes and a central resurgent dome hide its northern rim.*

*To visit Bandelier National Monument, turn left as directed by highway signs. Discussed in Chapter V, its story is an essential part of the Jemez Mountains story.*

West of Bandelier, the highway climbs gradually across the Bandelier tuff and the ring of older volcanic rock that formed the base of the Jemez volcano. It's remarkably easy to distinguish the central strongly welded unit of Bandelier tuff from the less firmly welded upper layer: Highway cuts in the welded zone are nearly vertical, but cuts in non-welded tuff, which slides easily, have been beveled to low angles.

The road climbs steeply west of Los Alamos. Coming over the ridge, we suddenly emerge into the Valles Caldera. The caldera is 14 miles across, but you can't see all of it; the view is blocked by the great dome of Redondo Peak, formed by resurgence of the floor of the caldera fairly soon after the great collapse. Around it are several smaller lava domes that later squeezed up through circular faults around the edge of the caldera; they conceal much of the real rim. The highway runs along the north slope of one of these domes.

We leave the caldera at milepost 40, and descend gradually through more Bandelier tuff exposed in large boulders and rugged pinnacles among the trees. Here again the distinction between tightly welded and unwelded tuff can be made on the basis of road-cuts. Black volcanic glass—obsidian—directly overlies some of the tuff; it is the material of one of the lava domes. Obsidian develops

where especially dry, high-silica lava cools without formation of mineral grains.

At the Jemez Canyon Overlook just west of milepost 29, a short nature walk provides views down into the canyon that drains the Valles Caldera. Before this canyon was cut, the caldera held a "crater lake." More obsidian, with white amygdules (gas bubble holes filled in with other minerals) is exposed just across the highway.

*The Valles Caldera resulted from volcanic explosions far larger than the Mt. St. Helen's 1980 eruption, followed by collapse of the volcano. Domes within the caldera formed later, as did lakebeds that floor the caldera.*, by permission, from Pages of Stone: Geology of Western National Parks and Monuments, 3: THE DESERT SOUTHWEST, by Halka Chronic (The Mountaineers, Seattle).

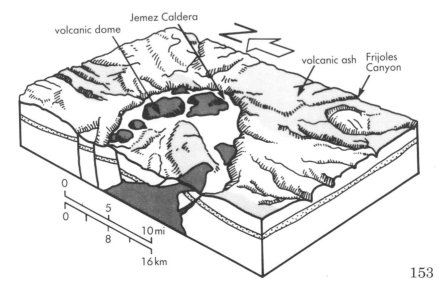

*Landslides in Jemez Canyon remind us that geology is an ongoing process.*

Leaving the parking area, the road drops into Jemez Canyon, passing more exposures of Bandelier tuff on the way. In the canyon depths are Tertiary volcanic rocks that underlie the Bandelier tuff and overlie Paleozoic sedimentary rocks.

The so-called "soda dam" just above Jemez Springs blocks the canyon and the Jemez River. Still enlarging, the dam developed where hot-spring water cooled and precipitated calcium carbonate—not sodium bicarbonate as its name suggests. In this entire area hot rocks are fairly near the surface—a legacy of the volcanic past—and groundwater heated by contact with them rises by convection to the surface.

The Bandelier tuff forms high cliffs east and west of Jemez Springs. Its three layers can be distinguished in many places, with conical

*The so-called "soda dam" in Jemez Canyon is not soda but calcium carbonate – travertine deposited by hot springs near the road. Jemez Creek (lower right) has tunneled under the dam.*

154

*"Tent rocks" in the Bandelier tuff decorate the high walls of Jemez Canyon. Each "tent" may represent a cone of hardened ash surrounding the escape route for volcanic fumes.*

"tent rocks" showing up in the middle layer. Note the very irregular surface on which the ash was deposited, full of valleys and ridges. The thickness of the ash here gives a feeling for the magnitude of the Jemez explosions.

Continuing down the canyon we find that red Triassic sandstone and shale gradually take over the scenery. These rocks—some of them crossbedded—dip southward off the Nacimiento Mountains west of Jemez Pueblo.

San Ysidro is surrounded by Triassic and Jurassic sedimentary rocks. Cuestas south of town are capped with Jurassic rocks containing thick beds of gypsum, some of it mined atop the cuestas. Gypsum forms when sea water or salty lake water evaporates. It is used for cement, plaster, wallboard, and other building materials.

*Red Mountain is striped with Triassic redbeds and older sedimentary rocks, some of them standing on edge. The mountain is a fault-edged syncline.*

155

San Ysidro is the site of a famous dinosaur "dig." In 1978-80 the bones of the large dinosaur *Camarasaurus* were excavated from Jurassic rocks near here. A *Camarasaurus* skeleton is on display at the New Mexico Museum of Natural History in Albuquerque. Another dinosaur, *Seismosaurus*, has also been discovered near San Ysidro.

The highway curves around the south end of the Nacimiento Mountains—one of the two southern tips of the Rocky Mountains—and then heads northward toward Cuba, traveling on and among Triassic and Jurassic rocks. The route follows the boundary between the Southern Rockies and the Plateau Country to the west. Flat-lying sedimentary rocks that characterize the plateau region bend up sharply toward the Nacimientos, and then break off completely where dark red Precambrian granite and metamorphic rock of the mountain core have been lifted along a large fault.

*A massive ledge of gray-white gypsum shows no sign of stratification. Erosion of fine sandstone below it gradually undermined the gypsum.*

Geologic patterns are complicated here by solution and distortion of salt and gypsum layers. Most of the salt has long since dissolved and washed away, contributing to collapse of sandstone and siltstone layers above it. The gypsum remains—thick white beds sandwiched between layers of sedimentary rock. In a way, gypsum is like ice: If you get enough of it in one place, under enough pressure, it flows. And in flowing it distorts rock layers above and around it. Overburden and

*Cabezon Peak, a large volcanic neck, preserves around its base some of the Cretaceous rocks that have been eroded away from the surrounding area.*

the rising of the mountain range undoubtedly provided the necessary pressure here.

Volcanic ash—here of Triassic age—has quite an influence on the scenery. It decomposes into clays that expand when they are wet and shrink as they dry, discouraging plant growth. The clays tend to form gray mats that in places blanket eroding hillsides.

Cabezon Peak, visible to the southwest from many parts of this highway, is the most prominent of several volcanic necks—lava-filled conduits of Tertiary volcanoes. Cabezon Peak preserves around its base some of the Cretaceous rocks that are eroded away from its immediate surroundings.

In several places near the highway, deep gullies cut into brown, silty stream deposits. Gullying may result from thinning of vegetation by grazing animals—horses, cattle, and sheep. Some of the gullies show looping meanders developed before gullying began, when the stream swung lazily across a nearly flat plain.

Mesas north and west of the highway, north of milepost 48, are capped with rocks of the Cretaceous Mesaverde group—a unit that shelters the famous cliff dwellings of Mesa Verde National Park in Colorado. These strata were deposited along the shores of the last sea to cover this area. They dip gently toward the northwest, into the San Juan Basin, a broad, oil-rich geologic sag that extends west to Arizona and north into Colorado. The picture is confused south of Cuba, however, as both Cretaceous and Jurassic rocks are overturned and appear to dip east.

North of Cuba the scenery changes again as we pass through some Tertiary sedimentary rocks eroded into colorful buttes, low hills, and badlands. Rounding San Pedro Mountain at the north end of the

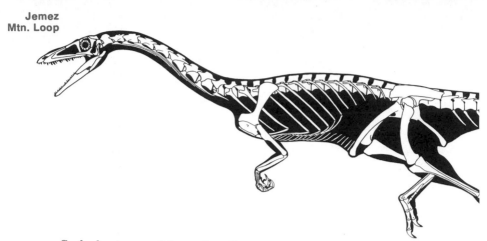

*Coelophysis, one of the earliest dinosaurs known, stood about as tall as a man. It ran swiftly on strong hind legs, and sported a frightening array of sharp*

Nacimiento Uplift, we cross the arch that separates the Chama Basin to the north from the San Juan Basin to the west. North of this divide are steeply tilted hogbacks of Mesozoic sedimentary rocks: Cretaceous, Jurassic, and Triassic units similar to those south and west of the mountains. Forested mountain slopes are surfaced with Pennsylvanian, Permian and Triassic sedimentary rocks. There are no Mississippian or older Paleozoic strata in this region; they were eroded away in Pennsylvanian time when this area was lifted to form one of the ranges of the ancestral Rocky Mountains. The skyline to the south is the north rim of the Valles Caldera.

Gypsum is common in Jurassic rocks at this end of the mountains, too. Thick gray-white beds of it are sandwiched between other Jurassic rock layers. It makes good carving rock, as it can be whittled with a pocket knife. Very fine-grained, compact gypsum is alabaster.

Near Coyote the highway descends through Triassic rocks into some of the Paleozoic strata. Ahead at milepost 27 is the eroded remnant of a basalt-capped plateau that once connected the San Pedro Mountains with more extensive plateaus to the north.

*Sunlit cliffs between Gallina and Coyote expose Triassic and Jurassic rocks capped with Cretaceous Dakota Sandstone.*

158

*teeth. A thousand specimens have been unearthed near Ghost Ranch.*
Reconstruction by Gregory Paul, copyright 1986.

Near mileposts 40 and 41 a grand panorama opens up—Abiquiu Reservoir backed with colorful Triassic redbeds and high cliffs of Jurassic sandstone. In Triassic rocks along the base of the cliffs, many dinosaur skeletons have been found, notably about 1000 specimens of a small 6- to 10-footer named *Coelophysis*—the oldest known dinosaur, recently designated "New Mexico State Fossil." At Ghost Ranch, a few miles north of the US 84 junction, the Ruth Hall Museum of Paleontology explains and displays some of the fossils found in this area.

Tertiary basalt caps the mesa south of Abiquiu. The Abiquiu Valley widens out east of town, and the river swings lazily in the meander pattern of low-gradient rivers. Gradually the river terraces along the Rio Grande come into view, with the Truchas Mountains, part of the Sangre de Cristo Range, 35 miles away across the Rio Grande Rift. A lava-covered mesa north of the highway, with rockslides developed from columnar-jointed basalt, is the southern tip of the Taos Plateau.

*Echo Amphitheatre, 3 miles north of Ghost Ranch, is a natural recess in Jurassic sandstone that when undermined tends to break away along arching joints. Cretaceous rocks top the mesa above.*

159

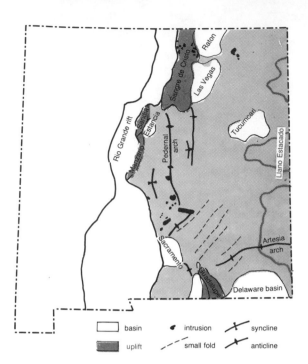

*The plains of eastern New Mexico slope toward Texas and Oklahoma, rising along a well defined escarpment that edges the Llano Estacado or Staked Plains.*

| | basin | ◆ | intrusion | ✕ | syncline |
|---|---|---|---|---|---|
| | uplift | - - - | small fold | ✕ | anticline |

# IV
# East of The Rift

From the Oklahoma and Texas border westward to the ranges that edge the Rio Grande Rift, the scenery of eastern New Mexico is that of broad plains and rolling hills interrupted here and there by low bluffs or shallow river valleys—part of the stable interior of the continent. In this part of the southern Great Plains, the geology is relatively simple—certainly far simpler than that of the Rio Grande Rift and the highlands of western New Mexico.

Here, flat-lying sedimentary rocks, ranging in age from Permian to Miocene, dominate. Because it resists erosion, Permian limestone—the San Andres limestone—provides a surface for much of the region, actually covering more acreage than any other rock unit in New Mexico. Other prominent units are red Triassic sandstones and siltstones, mostly of the Chinle formation, and the Cretaceous Dakota sandstone and Mancos shale, all well exposed in the northeastern part of the state. In places the plains are capped with patches of Miocene-Pliocene gravel, the Ogallala formation, continuous eastward far beyond New Mexico's border.

In all of this region there is very little in the way of geologic structure: a few faults, a few folds, and some very gentle warping of the sedimentary layers. Most of the relief is a product of erosion by the Pecos and Canadian rivers and their tributaries, and of collapse over caverns dissolved in limestone and gypsum. Mountains, where they occur, are nearly all volcanic. Well to the south, an unusual 250 million-year-old limestone reef lends a change of scene as it angles southward into Texas.

The Ogallala formation, formed of rock debris from the Southern Rockies and the rift-bordering ranges south of them, at one time extended right to the foot of those mountains. With reduction of rock debris as the mountains were worn down, steepening of streams by regional uplift in Pliocene time, and abundant rains of Pleistocene time, much of the Ogallala formation—the part nearest the mountains—was worn away. Its remnants along the state's eastern border are known as the Llano Estacado or Staked Plains.

Several of the valleys that lie close to the rift-bordering ranges held lakes in Pleistocene time, notably Lake Estancia east of the Sandia Mountains. The lakes left behind fine, silty lake deposits, wave-cut shorelines, and in many cases dune ridges along their eastern edges. Lake Estancia when full may have drained eastward to the Pecos River; when less than full it may have been salty, as were the playa lakes of many non-draining basins of southern New Mexico.

Dotted over the northern part of this region are hundreds of small volcanoes—cinder cones, shield volcanoes, and small lava domes. Some, Tertiary in age, are deeply eroded. Others are more recent—Pleistocene and Recent—and are scarcely changed by erosion. Northeast of the Canadian River, broad lava flows cover and protect part of the old Ogallala surface. North of Raton, remnants of older flows form high ramparts along the Colorado border.

One of the interesting aspects of the Canadian River, and in places of the Pecos as well, is that their courses have been to some extent governed by underground solution of salt and gypsum in Permian rocks. Flowing almost due south from Raton to Conchas Reservoir, the Canadian turns abruptly east, following the former margin of a Permian sea in which, as it evaporated, gypsum and salt were deposited. Near Santa Rosa the Pecos River clearly twists through a curving line of collapsed caverns, some of them now drowned by Santa Rosa Lake. The effectiveness of solution in creating collapse structures is well demonstrated by the many sinkholes that dot the region around Santa Rosa and other areas underlain by Permian rocks.

Several famous archeological sites exist in this part of New Mexico. Ancient campsites and spear points and other artifacts are found in association with bones of now extinct Pleistocene bisons and mammoths, and indicate that hunter-gatherers roamed these plains as early as 11,000 years ago, right at the tail end of the last Ice Age. More recent sites include villages built in the 12th to 14th centuries by ancestors of today's Pueblo Indians, groups with complex and interesting cultures based on agricultural crops of corn, beans, and squash. Many of these villages were thriving when the Spaniards arrived; some include Spanish churches and convents surrounded by the older villages.

Several of the highway descriptions in this chapter begin or end within the Rio Grande Rift, not because the eastern plains extend that far, but because population centers in New Mexico are concentrated along the Rio Grande, and make more convenient starting and stopping sites.

# I-25
## Colorado—Las Vegas
### 117 mi./188 km.

Interstate 25 enters New Mexico at the summit of Raton Pass, historically one of the main routes from the north and east, used by Indians, Spanish explorers and settlers, Union soldiers, and emigrants on the Bents Fort branch of the Santa Fe Trail. In 1879 a railroad was completed across the pass, and in 1922 the first highway—not much more than a narrow dirt track—was built.

On slopes above the pass, layers of Cretaceous and Tertiary sedimentary rocks are protected by a thick, resistant cap of lava flows 3.5 to 7.2 million years old. Most of these basalt flows came from volcanic centers near La Junta, Colorado, and spread in horizontal sheets well into New Mexico. They have gradually been whittled down by erosion; the cap on this high mesa is the largest remnant. Columnar joints mark some of the exposed lava cliffs.

Sedimentary rocks that underlie the basalt appear in highway cuts along Interstate 25: light brown sandstone, gray shale, and seams of coal—all parts of the Raton and Vermejo formations. These Cretaceous and Tertiary rock layers were deposited along the coast of a shallow sea, where mudflats and sandbars alternated with marshes and swamps. The layers vary in thickness, in places filling shallow channels. Elsewhere they thin and disappear, as we would expect along a land-sea boundary where beaches and lagoons are discontinuous. Coal from these rocks, visible in several roadcuts, has been mined on both sides of Raton Mesa; most of the mines are idle now.

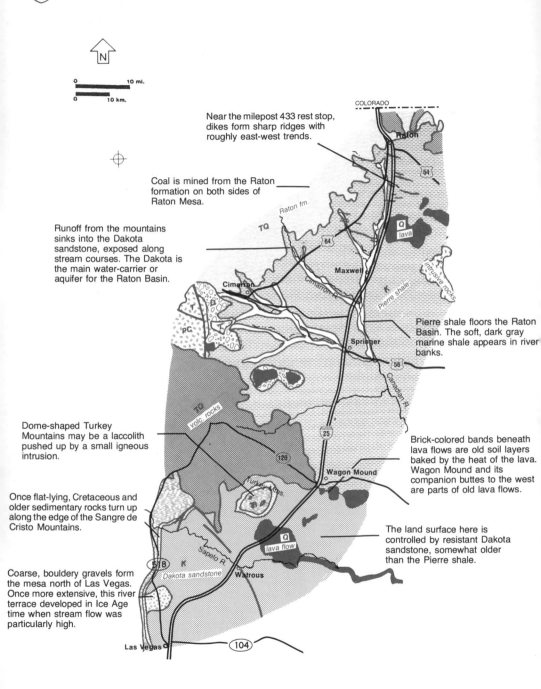

Near the milepost 433 rest stop, dikes form sharp ridges with roughly east-west trends.

Coal is mined from the Raton formation on both sides of Raton Mesa.

Runoff from the mountains sinks into the Dakota sandstone, exposed along stream courses. The Dakota is the main water-carrier or aquifer for the Raton Basin.

Dome-shaped Turkey Mountains may be a laccolith pushed up by a small igneous intrusion.

Once flat-lying, Cretaceous and older sedimentary rocks turn up along the edge of the Sangre de Cristo Mountains.

Coarse, bouldery gravels form the mesa north of Las Vegas. Once more extensive, this river terrace developed in Ice Age time when stream flow was particularly high.

Pierre shale floors the Raton Basin. The soft, dark gray marine shale appears in river banks.

Brick-colored bands beneath lava flows are old soil layers baked by the heat of the lava. Wagon Mound and its companion buttes to the west are parts of old lava flows.

The land surface here is controlled by resistant Dakota sandstone, somewhat older than the Pierre shale.

## Interstate 25
## COLORADO—LAS VEGAS

*The cliffs and slopes of Raton Mesa are composed of Cretaceous sandstone, shale, and coal. The mesa is capped with Tertiary sediments derived from the newly rising Rocky Mountains.*

The Cretaceous-Tertiary boundary lies within the Raton formation. This time boundary is receiving much attention now, in popular news as well as in geologic discussion, because it represents the time when all the dinosaurs and a large proportion of other types of animals and plants suddenly became extinct. The boundary is marked in many parts of the world by a thin layer of shale rich in iridium—an element rare on the Earth's surface but more abundant in meteorites. Many scientists now think that major meteorite showers caused the great extinction by throwing up, by explosive impact with the Earth, immense quantities of dust, thereby blotting out sunlight. The Earth's surface would have become Arctic-cold and dark perhaps for many months or years. Under these conditions most plants would perish and many animals would freeze or starve—the same scenario that scientists fear would result from a major nuclear war.

The town of Raton lies in the Raton Basin, near the head of the Canadian River's valley. To the west, Raton Mesa hides the Sangre de Cristo Mountains, southernmost range of the Rocky Mountains. To the east, the Great Plains are hidden by a broad anticline—the Sierra Grande Arch—and by many lava-capped mesas, buttes, and small volcanoes. An irregular Tertiary intrusion, Laughlin Peak, rises southeast of Raton as well.

When geologists speak of basins they are referring not to surface features but to rock layers beneath the surface. Under the surface here, sedimentary rocks bow downward between the Sangre de Cristo Mountains and the Sierra Grande Arch. Cretaceous sedimentary rocks that have been worn off both mountains and arch are still preserved in the basin. Notable among them are the Dakota sandstone, in a narrow line at the edge of the mountains, and the dark

165

gray Pierre shale, which floors the Canadian River's valley. A widespread, easily recognized unit, the Pierre shale derives its name from Pierre, South Dakota. It was deposited as mud on the floor of a fairly shallow sea, one of the last seas to cover the western interior of the continent. In places this shale contains beautifully preserved fossil shells of cephalopods (shell-bearing relatives of octopuses and squids) and other invertebrates, sharks' teeth, and skeletons of marine reptiles. The Pierre shale and mustard-yellow soils derived from it can be seen in the steep banks of the Canadian River.

Ranches and towns within the Raton Basin obtain water from the Dakota sandstone. This formation, a little older than the Pierre shale, is a beach and shore deposit. Porous and permeable like most beach sand, it serves as an aquifer (water-carrier) all up and down the east side of the Rockies. Water that sinks into it near the mountains flows eastward in the porous sandstone, held in by impermeable shale layers above and below. In the Raton Basin the Dakota sandstone also contains small amounts of oil, natural gas, helium, and carbon dioxide.

Near mileposts 436 and 435 the highway slices through two prominent dikes. The hard igneous rock of these dikes resists erosion, and they jut up in sharp ridges. On either side, well exposed in the highway cuts, the Pierre shale is baked and hardened by the heat of molten rock that formed the dikes.

The flat-topped mesa south and west of Springer is capped with Tertiary basalt that erupted from small volcanic centers not visible from the highway.

Between Springer and Wagon Mound, south of the Sierra Grande Arch, the highway reaches the edge of the Great Plains, which here as elsewhere are capped with Tertiary sandstone and conglomerate, the

*A dike south of Las Vegas baked and hardened the Pierre Shale on either side. The dike itself is about 10 feet thick; the baked shale adds 30 feet (15 on each side) to the resistant ridge-forming zone.*

166

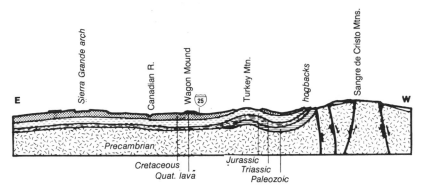

*Section across I-25 at Wagon Mound.*

Ogallala formation. In Pliocene time this formation, composed of gravel and sand washed from the Sangre de Cristos, extended right up to the mountains, covering all the Cretaceous sedimentary rocks we've just been looking at. However, the Canadian River has since carved its valley down through the Tertiary gravels and into underlying Cretaceous rocks. One gets a feeling here for the vastness of the plains, with their gently rolling topography and shallow stream courses.

The lava caps of Wagon Mound were a famous landmark on the Santa Fe Trail—six days by wagon to Santa Fe! There are two thick lava flows here, both Quaternary in age—a little younger than the lava flows of the mesa to the west. Soil under the flows was baked to a warm brick red by the heat of molten rock. The flows that form Wagon Mound also appear in patches of lava to the west and east; they came down a stream valley, filling it and preserving its form even after surrounding softer rock eroded away.

South of Wagon Mound the interstate travels on soil-covered Dakota sandstone—the aquifer of the Raton Basin. It, too, is a wide-

*Wagon Mound is part of two lava flows that for a time filled a valley. Ridges that bordered the valley have now eroded away.*

167

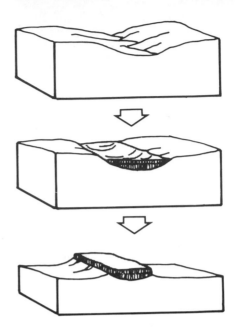

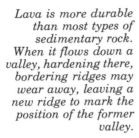

*Lava is more durable than most types of sedimentary rock. When it flows down a valley, hardening there, bordering ridges may wear away, leaving a new ridge to mark the position of the former valley.*

spread formation, extending north into Wyoming and the Dakotas. Visible to the southwest is a small symmetrical dome of Triassic and Jurassic rocks, dark with trees—the Turkey Mountains. This dome may be a laccolith, a small intrusion that squeezed between sedimentary layers, doming up those above. Some of the domed-up Mesozoic rocks have eroded from the summit, leaving a hogback ring of Dakota sandstone and a "racetrack" valley of soft Jurassic rock around the little mountain.

Mountains farther west, coming into view south of Fort Union, are again parts of the long chain of the Sangre de Cristos. As Precambrian rocks were faulted upward to form this range, sedimentary layers bent up abruptly along its east edge; they now form hogbacks along the mountain front.

South of Watrous the highway draws closer to the mountains. We're nearing the southern end of the Rockies, which extend as the continent's backbone from New Mexico to Alaska. Close to Las Vegas the faults that edge this side of the range peter out, although those on the west side continue southward as the margin of the Rio Grande Rift.

A flat-topped mesa just north of Las Vegas is a remnant of a Pleistocene alluvial fan. Erosion of the mountains reached all-time highs during Pleistocene time, when continued uplift of these mountains was coupled with heavy precipitation of the ice ages. Torrential streams flooding from the mountains must have built many such fans; few still exist.

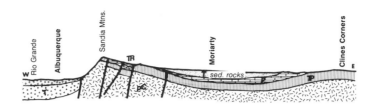

*Section along I-40 Albuquerque – Clines Corners*

# I-40
# Albuquerque—Clines Corners
### 61 mi./98 km.

Leaving Albuquerque, the route climbs across large alluvial fans deposited, mostly in Pleistocene time, by Tijeras and Embudo creeks. The Sandia Mountains rise about 4000 feet above these fans, 5000 feet above downtown Albuquerque. Their steep, rugged western face is an eroded fault scarp of Precambrian Sandia granite topped with thin layers of Mississippian and Pennsylvanian sedimentary rocks. The Precambrian-Paleozoic contact, near the summit of the range, has its counterpart 20,000 feet down below Albuquerque; total movement along this part of the rift fault therefore reaches some 25,000 feet—about 5 miles. Low fault scarps across alluvial fans at the base of the mountains—not visible from the highway—show that the faults have moved fairly recently. And they will probably move again, to an accompaniment of earthquakes, to which Albuquerque is no stranger.

The Sandia Mountains are a single very large fault block raised highest on its western side. We can expect to see younger and younger rocks as we drive eastward through the tilted fault block.

At the head of the Tijeras Creek alluvial fan, the route enters Tijeras Canyon, where large roadcuts vividly portray the rock sequence. Near the canyon entrance the Sandia granite weathers into steep, angular crags and pinnacles. In some highway cuts the normally blocky, jointed granite has clearly weathered, while even below the natural surface, into rounded residual boulders separated by masses of coarse sand composed of quartz and feldspar crystals

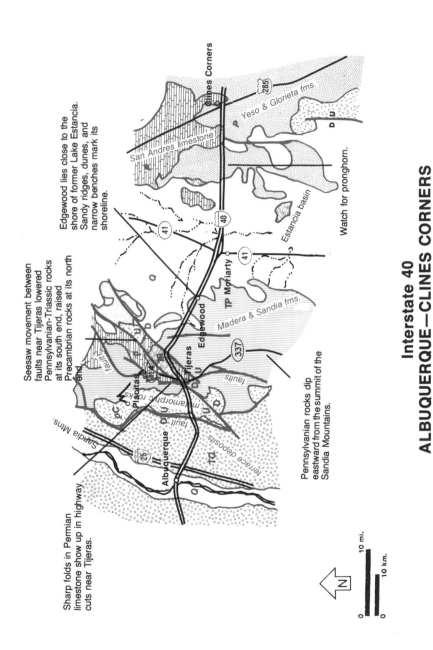

Interstate 40
## ALBUQUERQUE—CLINES CORNERS

Seesaw movement between faults near Tijeras lowered Pennsylvanian-Triassic rocks at its south end, raised Precambrian rocks at its north end.

Sharp folds in Permian limestone show up in highway cuts near Tijeras.

Edgewood lies close to the shore of former Lake Estancia. Sandy ridges, dunes, and narrow benches mark its shoreline.

Watch for pronghorn.

San Andres limestone

Yeso & Glorieta fms.

Clines Corners

285

Estancia basin

41

40

41

Moriarty

Edgewood

Madera & Sandia fms.

337

Tijeras

metamorphic rocks

faults

Placitas

Albuquerque fault

Sandia Mtns.

terrace deposits

25

Pennsylvanian rocks dip eastward from the summit of the Sandia Mountains.

N

10 mi.

10 km.

from the granite. Such boulders are the result of physical and chemical weathering along joints, with weathering attacking angles and corners from two or more sides at once. Watch for dikes in the granite roadcuts.

Tijeras Canyon developed in weakened rock along the Tijeras fault. South of Tijeras Canyon, most of the Precambrian rocks are metamorphic. Some north of the highway are, too: Steep reddish slopes between mileposts 171 and 172 are composed of Tijeras gneiss, a granite-like metamorphic rock chiefly recognized, at highway speeds anyway, by its color. Banding that usually characterizes gneiss is poorly developed in this rock, though it is cut by many dikes. The Tijeras gneiss has been dated as older than the Sandia granite, around 1.4 billion years old. In places it grades into the granite as if the intruding rock had melted its way through the gneiss, partly turning it into granite, too.

East of milepost 172 both roadcuts and canyon slopes take on a greenish hue derived from the Tijeras Canyon greenstone, a rock unit made up of metamorphosed lava flows. These rocks, also Precambrian in age, have dropped to roadside level between two faults. The narrow block in which they occur formed in seesaw fashion—one end down, the other up. Rocks near the faults are bent in little folds and broken by many smaller faults. They contain gold-bearing quartz veins as well.

The Precambrian greenstone directly underlies Pennsylvanian gray shale and limestone, which are also part of the down-dropped sliver. These strata contain many invertebrate fossils, including brachiopods, corals, bryozoans, clams, snails, trilobites, and button-like sections of crinoid stems. Mudstone layers in the formation locally contain fossil ferns and impressions of leaves of other land plants. In combination these two groups of fossils—marine shellfish and land plants— suggest that these rocks were deposited in shallow bays and swampy estuaries, near a gently sloping coastline where small changes in sea level submerged sizeable amounts of land.

Near Tijeras some younger sedimentary rocks appear—dark red-

*Tilted Pennsylvanian sedimentary rocks dip eastward off the Sandia Mountains.*

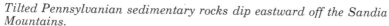

dish brown sandstone and shale, "redbeds" in geologic lingo, of Triassic age. Many of the individual sandstone layers are crossbedded, with the short, straight crossbeds that indicate that they were deposited by moving water. In places the red rocks contain petrified wood, leaf impressions, and footprints of Permian reptiles—all of which demonstrate that the unit was deposited above sea level, probably on a floodplain or delta where streams spread out on a relatively flat surface. The color of the rocks is due to tiny particles of iron oxide that coat the sand grains or make up part of the clay of the mudstone. The canyon widens out in these less resistant sedimentary rocks.

East of Tijeras there are views of the "back" side of the Sandias, with tilted Paleozoic sedimentary rocks forming the mountain slope. Near milepost 175 a small fold, the Tijeras anticline, surfaced with Mesozoic rocks, shows up north of the highway.

At the small community of Edgewood the highway enters the Estancia Basin, with the snow-capped Sangre de Cristo Mountains, southern tip of the Rocky Mountains, in the distance to the north. Edgewood obtains its water from Pennsylvanian rocks of the Madera group, replenished from the east slope of the Sandias. This unit measures about 1300-1400 feet thick in the mountains, but thickens eastward to more than 2000 feet in the Edgewood area. Water travels through joints and solution cavities in the limestone layers, and in some places through porous sandstone units. The water contains quite a bit of calcium carbonate, popularly but incorrectly called "lime," as well as high levels of magnesium, sodium, potassium, and fluoride.

In the Estancia Basin the Madera formation is overlain by the Permian Abo formation or by Quaternary stream and lake deposits which partly fill the basin. Fine, dark red shales in the Abo formation are relatively impermeable, and help to confine water in underlying layers. Similarly, fine lake sediments within the Estancia Basin confine water in much coarser stream deposits. The basin has no outlet. In Pleistocene time, with Ice Age increases in precipitation, it was the site of a large lake-perhaps a salty one part of the time, but fresh when and if it found outlets through the many underground passages in surrounding limestones. The lake's beach ridges, narrow benches cut by lapping waves, appear clearly on aerial photographs.

East of the Estancia Basin the highway climbs gradually through Pennsylvanian and Permian sedimentary rocks. Watch for red shales and sandstone of the Abo formation, and east of them gray San Andres limestone. To the south are the Pedernal Hills, upfaulted blocks of Precambrian granite. The relationship between these granite hills and surrounding sedimentary rocks tells us that they first pushed upward in Pennsylvanian time, as the southern outpost of the Ancestral Rocky Mountains.

Small, shallow, usually circular basins in this area are sinks collapsed over caverns dissolved by groundwater in the San Andres limestone and underlying formations.

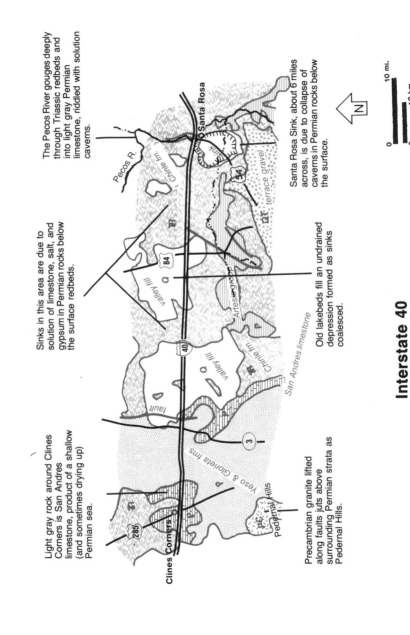

The Pecos River gouges deeply through Triassic redbeds and into light gray Permian limestone, riddled with solution caverns.

Santa Rosa Sink, about 6 miles across, is due to collapse of caverns in Permian rocks below the surface.

Sinks in this area are due to solution of limestone, salt, and gypsum in Permian rocks below the surface redbeds.

Old lakebeds fill an undrained depression formed as sinks coalesced.

Light gray rock around Clines Corners is San Andres limestone, product of a shallow (and sometimes drying up) Permian sea.

Precambrian granite lifted along faults juts above surrounding Permian strata as Pedernal Hills.

Pecos R.

Santa Rosa

Chinle fm.

terrace gravel

Qt

valley fill

Q

84

40

valley fill

Q

Chinle fm.

San Andres limestone

Artesia gp.

3

Yeso & Glorieta fms

Pedernal Hills

pC

285

Clines Corners

fault

54

N

0          10 mi.

0          10 km.

**Interstate 40**
## CLINES CORNERS—SANTA ROSA

# I-40
# Clines Corners—Santa Rosa
### 54 mi./87 km.

The pale gray rock around Clines Corners is part of the Permian San Andres limestone, a marine limestone that covers more of New Mexico than any other single formation. The San Andres is thin here, but thickens southward, which was seaward in Permian time. We stay on its surface—a grass-covered tableland spotted with juniper trees—for several miles, passing or crossing a number of undrained, silt-floored depressions, sinks formed by collapse in and below the limestone.

Underground, the San Andres formation is riddled with solution passages and caverns. This is the formation that includes the thick reef limestone that hosts Carlsbad Caverns 200 miles to the south. Rock units beneath it, notably the Yeso formation, contain lots of gypsum and once included thick beds of salt. Since both salt and gypsum are even more soluble than limestone, many of the sinks result from solution of these minerals.

Mesas visible to the north are capped with light tan, crossbedded Glorieta sandstone, a fine, even-grained quartz sandstone formed on a Permian beach just before the San Andres limestone was deposited. The unit, often considered to be the base of the San Andres formation, is quite near the highway at milepost 225.

In the distance to the south from milepost 219, the Pedernal Hills are the remains of a Pennsylvanian mountain range, part of the so-called Ancestral Rocky Mountains. They are made up of Precambrian igneous and metamorphic rock, and were once topped with Paleozoic sedimentary rocks. Partly edged with faults, they are really the southern tip of the Sangre de Cristo uplift, and therefore of the Rockies, though for most purposes the Rockies are considered to end near Santa Fe.

Near mileposts 228 and 230, roadcuts expose red and yellowish green shale and whitish limestone of the Yeso formation, with the Glorieta sandstone on top. Veins of gypsum are visible in this rock. Glorieta sandstone appears again east of the Yeso roadcuts and on a mesa to the south.

Near milepost 237 the highway crosses a high, level surface—a small outpost of the Great Plains, which in late Tertiary time extended all the way west to the mountains. Here as elsewhere the Great Plains are surfaced with the Ogallala formation, made up of Miocene-Pliocene gravel washed eastward from the various ranges of the Rocky Mountains and the Basin and Range areas south of them. The Ogallala formation is thin here, and along this highway it is mostly covered with silty soil or with hummocky sand dunes left over from Pleistocene time, now stabilized by vegetation.

About 12 miles east of Clines Corners we encounter colorful Triassic sandstone and siltstone of the Chinle formation. Below them, but also Triassic, is the Santa Rosa sandstone. These rocks are gently warped, forming low anticlines and synclines. The harder sandstone beds weather into ledges, the softer siltstone layers into slopes.

One of the few places along this highway where you can get a close look at some of these rocks is near the rest area at milepost 251, where part of the Santa Rosa formation is exposed on the walls of a collapse valley visible from the top of a small hill within the rest area. The same rocks can be seen in a nearby highway cut and edging the mesa to the southeast. Here the Santa Rosa formation is mostly light brown or gray sandstone with brightly sparkling feldspar grains. Between the sandstone layers the formation also includes bands of dark reddish brown mudstone, as well as some conglomerate. Weathered surfaces are generally dark. Watch for more exposures of these rocks in roadcuts between the rest stop and the junction with US 84.

Both the Santa Rosa and Chinle formations contain petrified wood and other fossil plants, and invertebrate and vertebrate fossils. Notable among the last are phytosaurs, crocodile-like reptiles that inhabited the floodplains of Triassic rivers.

East of the US 84 junction the highway crosses more Triassic rocks, partly hidden by silty Pleistocene soils and sand. This whole area is pocked with sinks caused by solution of limestone in the San Andres formation or salt and gypsum in the Yeso formation, both beneath the surface here. Topography developing because of solution of underlying rocks—whether limestone, gypsum, or salt—is called karst—

*The Blue Hole at Santa Rosa is a sink formed by partial collapse of an underground cavern. Scuba divers explore its crystal depths.*

named for a cavernous limestone region in Yugoslavia. Here the karst is not as well developed as in wetter climates, but solution has caused enough sinks and undrained depressions to allow use of the term. Both the Pecos River and its tributaries in places follow collapse valleys. Just west of Santa Rosa the highway drops rather suddenly into the Santa Rosa Sink, a large flat-floored collapse area more or less circular in outline and some six miles across.

Interstate 40 crosses the Pecos River at Santa Rosa. About ten miles south of the town, in 1541, Coronado and his little army, searching for the legendary seven cities of gold, built a log bridge to cross the Pecos.

There are several sinks right in the town of Santa Rosa, among them the Blue Hole with a diameter of 60 feet, a depth of 81 feet, and a flow of 3000 gallons per minute. Its extremely clear 61-degree water is a favorite of scuba divers.

A few miles north of town, near Santa Rosa Lake, the Santa Rosa sandstone includes tar sands containing an estimated 90 million barrels of oil. The tar sands were mined for road material in the 1930s, but are not being worked at present. Before it thickened to tar, the oil migrated upward from Pennsylvanian and Permian source rocks through fractures and faults created when the Santa Rosa sandstone collapsed into solution cavities.

Santa Rosa Dam, on the Pecos River upstream from the town, serves for conservation of irrigation water, flood control, and sediment control. There are many good exposures of the Santa Rosa formation and overlying red rocks near the road to the dam, particularly in the walls of Pecos Canyon below the dam, where the river has cut down through nearly all of the Santa Rosa formation. The San Andres limestone is exposed in spillway cuts. Some sandstone surfaces display ripple marks and tracks and trails of aquatic animals.

Archeological studies before the dam was built identified many sites—Indian, Spanish, and American—ranging in age from 5000 B.C. to 1870 A.D.

*Thick deposits of caliche near Cuervo are dug for road materials.*
R.L. Griggs photo courtesy of U.S. Geological Survey.

# I-40
# Santa Rosa—Texas
## 100 mi./161 km.

East of Santa Rosa, Sunshine Mesa comes into view to the south, its red slopes capped with caliche, an impure mixture of calcium carbonate and soil developed since Pleistocene time. Caliche develops in arid climates where soil water containing dissolved calcium carbonate is drawn to the surface and evaporated, leaving crusty deposits of calcium carbonate in the soil. On this mesa caliche deposits are thick enough to be mined for road material; look for small mine dumps near the edge of the mesa.

Below the hard caliche cap are red sandstone, siltstone, and conglomerate of the Chinle formation, the Triassic rock that overlies the Santa Rosa formation. In places these red rocks contain fossil plants and amphibian and reptile bones, as well as a few invertebrate animals such as clams and snails. Study of the Triassic rocks and their fossils shows that the rocks formed where sand and mud were deposited by streams entering a large freshwater lake. Shallow stream estuaries and marshy lakeshores supported clusters of rushes and other marsh plants. Large, sluggish, flat-headed amphibians and more agile crocodile-like phytosaurs inhabited the mud flats and splashed through the water to capture fish. Small clams and snails burrowed in the mud or clung to the rushes. Ferns and a few trees

179

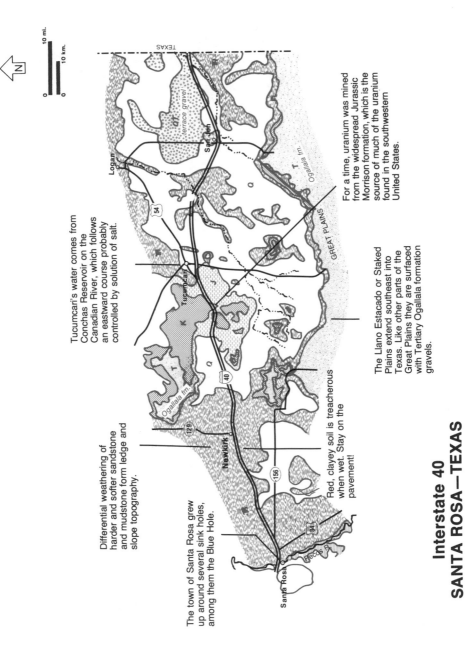

Tucumcari's water comes from Conchas Reservoir on the Canadian River, which follows an eastward course probably controlled by solution of salt.

Differential weathering of harder and softer sandstone and mudstone form ledge and slope topography.

The town of Santa Rosa grew up around several sink holes, among them the Blue Hole.

For a time, uranium was mined from the widespread Jurassic Morrison formation, which is the source of much of the uranium found in the southwestern United States.

The Llano Estacado or Staked Plains extend southeast into Texas. Like other parts of the Great Plains they are surfaced with Tertiary Ogallala formation gravels.

Red, clayey soil is treacherous when wet. Stay on the pavement!

**Interstate 40**
**SANTA ROSA—TEXAS**

grew where hills rose above the general floodplain surface. Far to the west and northwest were the remains of the Ancestral Rocky Mountains, by Triassic time eroded to low, rolling hills.

North of Newkirk and 4000 to 8000 feet below the surface, the Santa Rosa formation contains an estimated 39 million barrels of heavy oil. Because of difficulties in extracting this thick oil, it is not being produced. However, it can be considered part of the nation's oil reserve, obtainable when the price of oil exceeds the cost of extracting it.

Both northeast and southeast of Newkirk, and visible from the rest area east of town, are higher mesas capped with a much younger rock unit—the Miocene-Pliocene Ogallala formation, caprock of the Great Plains. These mesas are isolated or nearly isolated remnants of the Great Plains, which formerly extended without interruption all the way to the Rocky Mountains, source of their gravel and sand. As you drive eastward you'll see that the southern of these two remnants does connect up with the main Great Plains surface, here called the Llano Estacado or Staked Plains.

The caliche-hardened Ogallala formation on the Great Plains surface overlies Jurassic rocks, which in turn rest on the Triassic rocks described above. Since the highway will rise onto Jurassic rocks east of Montoya, let's take a moment to discuss them:

The lowest Jurassic formation exposed here is the Entrada sandstone, a cliff-former made up of thick layers of crossbedded, round-grained sandstone formed from sand dunes. The Entrada is beveled at the top by limestone and gypsum of the Todilto formation, deposited in a large bay isolated from the sea, or in a salt-water lake.

An interesting aspect of some of the Todilto limestones is that they accumulated in thin varves, each varve consisting of three paper-thin layers, one of limestone, one of dark organic material, and one of fine

181

clay. Varves are produced by yearly variations in runoff, water temperature, and abundance of lake organisms. Accurate counting of the paper-thin varves, as if they were tree rings, in the lower part of the Todilto formation gives us the length of time, in years, involved in its accumulation: 14,000 years. The upper part of the formation contains a good deal of gypsum and is less regularly varved. Fossils are rare, though some of the varved limestone contains fossil fishes.

The Morrison formation, above the Todilto formation but still Jurassic, is composed of soft, easily eroded green and red shale and siltstone, with a few sandstone and conglomerate beds, all deposited in a river floodplain environment. The shales include several clay minerals known to be derived from volcanic ash, as well as uranium. There are some famous dinosaur sites in the Morrison formation in Colorado and Utah; here in northeastern New Mexico the unit contains a few dinosaur bones and some fragments of petrified wood.

Near milepost 319 the highway climbs onto these Jurassic units. Variegated red, yellow, and green shales of the Morrison formation are easily recognized even at highway speeds. A few miles farther east the road climbs again, onto the Dakota sandstone, a thin but very widespread beach and near-shore sandstone that signals the return of the sea in Cretaceous time. Near milepost 329 a large highway cut slices neatly through the contact between the Jurassic and Cretaceous formations, and reveals a channel in the top of the Morrison formation, filled in with orange Dakota sandstone.

Approaching Tucumcari the highway goes down through these Cretaceous, Jurassic, and Triassic rocks again, into the wide valley of the Canadian River. In Pliocene time, when erosion was spurred by uplift of this entire region, this river broke through the Ogallala formation caprock of the Great Plains to form what is known as the

*Paper-thin layers in the Todilto Limestone are varves, annual increments of very fine sediments.* R.L. Griggs photo courtesy of U.S. Geological Survey.

182

"Canadian Breaks"—a broad indentation in the west margin of the High Plains. Within the dent, terraces developed in response to cyclic climate changes of the Ice Age.

All around Tucumcari are salmon-colored soils derived from Triassic rocks—soils that with the addition of water from the Canadian River, about six miles north of the town, support prosperous farms. The highway continues eastward over these salmon soils, with portions of the Great Plains escarpment in the distance to the south and north. White caliche marks low spots and shows where water has evaporated.

Because the irregular edge of the Llano Estacado forces winds to rise and therefore to lose some of their carrying power, wind-deposited sand dunes are common along this stretch of the highway and elsewhere along the edge of the Llano Estacado. The dune fields, developed during Pleistocene time, are now pretty well stabilized by growth of vegetation.

Where streams have cut through the pink soil, Triassic rocks are exposed. Between mileposts 346 and 347, for instance, Revuelto Creek has incised a particularly deep canyon. A few miles south of the highway, along this creek, are fossil vertebrate sites from which skulls and bones of phytosaurs, fish, amphibians, and a small meat-eating dinosaur have been recovered by teams from Yale University and the Universities of Colorado and New Mexico. In searching for such sites, paleontologists work their way upstream watching for bone fragments, then continue upstream to find the sources of the fragments. Unfortunately, no complete skeletons have been found at the Revuelto Creek localities.

The highway continues to converge with the edge of the Llano Estacado. East of San Jon it rises onto a surface covered with redeposited gravel derived from the Plains surface. As we approach the Texas border we get a good view of this remarkably flat surface, which has remained virtually unchanged, except for erosion nibbling at its edges, for several million years. Reinforced with caliche, the caprock absorbs water readily, thereby preventing development of streams strong enough to erode channels into it. Its porosity serves also to provide abundant well water for ranches and farms on its surface. The Great Plains have counterparts, both scenically and geologically, in the steppes of Russia and the pampas of South America.

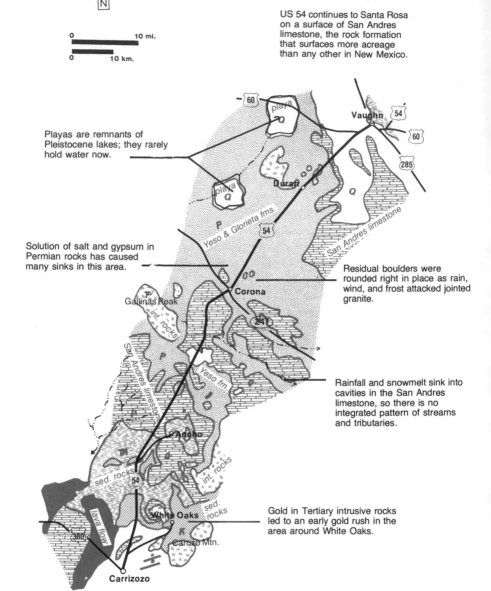

US 54 continues to Santa Rosa on a surface of San Andres limestone, the rock formation that surfaces more acreage than any other in New Mexico.

Playas are remnants of Pleistocene lakes; they rarely hold water now.

Solution of salt and gypsum in Permian rocks has caused many sinks in this area.

Residual boulders were rounded right in place as rain, wind, and frost attacked jointed granite.

Rainfall and snowmelt sink into cavities in the San Andres limestone, so there is no integrated pattern of streams and tributaries.

Gold in Tertiary intrusive rocks led to an early gold rush in the area around White Oaks.

# US 54
# CARRIZOZO—VAUGHN

# US 54
# Carrizozo—Vaughn
### 82 mi./132 km.

From Carrizozo, as well as from US 54 north of town, black lava of the Valley of Fire lava flow can be seen to the west, with Little Black Peak, source of the lava, near its northern end.

A short distance northeast of Carrizozo, hidden between Carrizo Mountain and the Jicarilla Mountains—a cluster of Tertiary intrusions—is the town of White Oaks, a gold-mining boom town of the 1880s. Placer gold was discovered there in 1850, but mining was sporadic because of isolation, Indians, and the Civil War. The town didn't really develop until thirty years later, when gold was mined from intrusive rocks west of town. Around 150,000 ounces of gold were mined from this district. But by the turn of the century White Oaks was a ghost town, its riches forgotten, some of its more elegant homes moved to Carrizozo.

The highway proceeds north across a large alluvial fan containing gravel from the Jicarilla Mountains—a well defined, probably Pleistocene fan with its lowest watercourses marked by lines of desert shrubbery. Fans from adjacent canyons merge with it to form an alluvial apron around the range as a whole. Roadcuts south of milepost 136 show that part of the fan is underlain by dark basalt, whitened with caliche near the surface. Elsewhere the fan is underlain by purplish Jurassic shale.

The Jicarillas are a cluster of irregularly shaped laccoliths, their igneous mass doming up overlying sedimentary rocks that range from Permian sandstone and limestone to Cretaceous sandstone and shale. These rock units have eroded off the central core of the mountains, but appear in concentric rings around the central intrusion. We come on some of the Cretaceous rock—notably the ridge-forming

*A pitlike sink marks the collapsed ceiling of an underground cavern.*

Dakota sandstone—near milepost 137. Farther north we pass through Jurassic shales and onto deep red Triassic siltstone and sandstone, in which we remain for some distance. Hills on either side of the road are bright with these colorful rocks.

Still farther north the highway crosses a roller-coaster surface of pale gray Permian San Andres limestone. The terrain is dotted with sink holes where the limestone has collapsed into underground caverns. North of milepost 152 we converge with light-colored hills of Yeso formation, a Permian rock unit slightly older than the San Andres formation.

In this area the contact between these two units is marked by a thick yellowish or reddish soil layer developed before the San Andres sea advanced across the Yeso formation. The Yeso formation contains both gypsum and bentonite, the latter a group of clay minerals commonly derived from volcanic ash. Gypsum is relatively soluble, and bentonite tends to expand when it gets wet and shrink again when it dries. As a result, portions of the Yeso formation slough off easily, and in places work slowly downslope as gray, surface-covering mats. The formation originally contained thick layers of salt, much more easily dissolved than gypsum; solution of the salt is responsible for many sinks in this area.

Another fairly large Tertiary igneous intrusion juts up as the Gallinas Mountains, west of milepost 162.

From the overpass north of Corona several rounded rocky outcrops to the north look like the Loch Ness Monster surfacing through waves of juniper trees. These hills are isolated outcrops of coarse-grained Precambrian granite, exposed here where Permian sedimentary rocks that normally cover them have worn away. They display a characteristic weathering pattern of granite (and some other rocks): large boulders formed by gradual rounding of sharp-edged blocks.

186

No Cambrian, Ordovician, Silurian, Devonian, Mississippian, or Pennsylvanian rocks exist here; Permian rocks of the Yeso formation rest directly on Precambrian granite. If older Paleozoic rocks ever covered this area, they were worn away in late Pennsylvanian and early Permian time, as uplift raised a north-south ridge stretching from here to Colorado, the southernmost part of the Ancestral Rocky Mountains.

North of Corona much of the surface is patched with Pleistocene gravel thinly covering Permian strata. The surface strongly resembles that of the Great Plains—fertile, rolling hills with summits all at about the same elevation. The true Great Plains, capped with Miocene gravel, are farther east.

Hills near Duran are held up by sills, sheetlike horizontal intrusions that forced their way between layers of Permian rocks. Sinks between Duran and Vaughn were caused by solution of gypsum in the Yeso formation, as is a collapse valley just north of milepost 197.

Collapse occurs, too, in limestone of the San Andres formation. Limestone dissolves much less easily than gypsum or salt, and only in slightly acid water. However, rain picks up enough carbon dioxide from the atmosphere and soil to become weakly acid, and over long periods of time can carve out extensive cavern systems. Northeast of Vaughn are many sinks formed by limestone solution, most of them shallow. Some expose the blocky tan limestone.

A few of these sinks hold water. There is no regular drainage pattern on the limestone surface—no well-defined streams or stream valleys. When snow melts or storms occur, the runoff flows toward individual sinks, and there disappears underground, to etch more caverns in the limestone or to enlarge those already there.

The solubility of limestone is further illustrated by whitish caliche on the surface and in roadcuts. Caliche develops where moisture that has entered the limestone and absorbed calcium carbonate from it is drawn to the surface and evaporated. Only the water evaporates; the dissolved minerals remain in or on the soil surface.

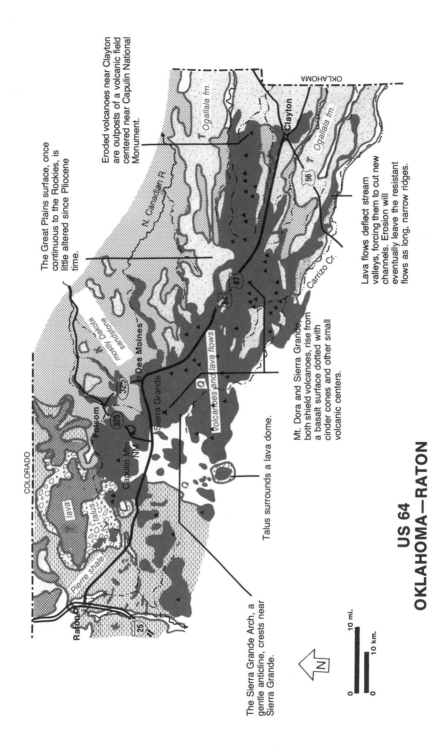

Eroded volcanoes near Clayton are outposts of a volcanic field centered near Capulin National Monument.

The Great Plains surface, once continuous to the Rockies, is little altered since Pliocene time.

Lava flows deflect stream valleys, forcing them to cut new channels. Erosion will eventually leave the resistant flows as long, narrow ridges.

Mt. Dora and Sierra Grande, both shield volcanoes, rise from a basalt surface dotted with cinder cones and other small volcanic centers.

Talus surrounds a lava dome.

The Sierra Grande Arch, a gentle anticline, crests near Sierra Grande.

**US 64**
**OKLAHOMA—RATON**

*Rabbit Ear Mountain near Clayton is a remnant of a Tertiary volcano. Terracelike levels below it are lava flows. The windmill pumps water from underlying Ogallala formation gravels that elsewhere form the surface of the Great Plains.*

# US 64
# Oklahoma—Raton
## 86 mi./138 km.

This highway enters New Mexico on the Great Plains surface, a gently rolling upland of Miocene gravel and sand that has existed with very little change since Pliocene time. At Clayton Lake State Park about 12 miles north of Clayton, more than 500 dinosaur footprints are concentrated in a two-acre exposure of Cretaceous sandstone. Near Clayton, Rabbit Ear Mountain punctuates the monotony of the landscape. The eroded remains of a volcano, the mountain is surrounded by long lava flows that stretch eastward toward the state line. Rabbit Ear Mountain was a landmark on the Cimmaron cutoff of the Santa Fe Trail, which approximately paralleled the present US 64. Hundred-year-old wagon ruts are still apparent in places—a measure of the slow erosion rate in this semi-arid land where the gravelly surface of the Plains absorbs rain and snowmelt. The scenery itself has changed little since the ruts were formed. Subtract the railroad, highway, phone poles, and fence, and let the grass grow tall, and you have this region as it was when Conestoga wagons rolled their way across it.

Clayton is up on a lava flow surface. West of town the highway travels for many miles on the crest of a higher tongue of lava, now well grassed over. The fluid basaltic lava originally flowed down the valley of an east-flowing stream, displacing the stream. Later, as ridges that edged the valley eroded away, the lava flow remained as a

189

resistant ridge—a reversal of topography fairly common in north-eastern New Mexico. A glance at the geologic map reveals that many other flows, now ridges, parallel this one and have a similar history.

Some of the dark gray basalt blocks along the railroad and fence-line are coated with white caliche, a mixture of calcium carbonate and other soluble minerals deposited as mineral-carrying moisture, drawn upward through joints in the rock, evaporated.

Occasional small depressions in the lava surface mark collapses of lava tunnels formed where molten lava flowed out from under its own hardened crust.

The dome-shaped mountain northwest of milepost 367, with radio relay towers, is Mt. Dora, a shield volcano formed by lava slightly less fluid than that in the flows we've been seeing. West and northwest of Mt. Dora are other volcanoes, some steep-sided and partly eroded, others with the classic low profiles of shield volcanoes. Most have craters, though some of these are breached by erosion. Seen from above or on a topographic map, streams draining such shield volcanoes show a characteristic radial pattern.

*Sierra Grande, a shield volcano, can be seen from the summit of Capulin Mountain. The lava flow at the bottom of this picture comes from Capulin.*

At milepost 357, Sierra Grande (8720 feet) is straight ahead. One of New Mexico's largest volcanoes, it rises above a long anticline known as the Sierra Grande Arch, which separates the area in which we've been traveling from the Raton Basin, a geologic sag that flanks the Sangre de Cristo Mountains.

The highway curves north around Sierra Grande, through the town of Des Moines. Near here, archeologists have found many skillfully made stone spear points similar to those originally discovered at Folsom, about six miles to the north. The Folsom points were produced by hunter-gatherers that camped in this region before more advanced Indian cultures developed.

Capulin Mountain, a cone-shaped hill with a road spiraling up it, comes into view west of Des Moines. Centerpiece of a national monument, the mountain is a cinder cone formed as fragments of foamy lava, hurled into the air from a volcanic vent, fell back to the ground. Lava flows around its base came from low on the west side of the cone. One flow extends right to the highway, presenting a rough, irregular surface. When such lava moves, it seems to roll along like a military tank, laying down a "tread" of cooled and broken pieces that fall from the front of the flow, then slowly pushing forward onto these pieces as it carries more cooled and broken blocks forward on its upper surface.

The rounded volcano almost due south of Capulin is a lava dome formed of particularly thick, viscous lava, the consistency of bread dough, that pushed up through older flows.

Take a good look at the long lava-capped mesa southwest of Capulin. As you get farther west you'll discover that it is much narrower than it is long—another valley-following flow left as a ridge by erosion of softer rocks on either side.

From milepost 314 west, the sawtooth profile of the Sangre de Cristo Mountains marks the western skyline. These mountains are the southernmost part of the Rocky Mountains. Below their snowy crest is Raton Mesa, walled with soft Cretaceous marine shales and capped with Cretaceous-Tertiary Raton formation, a sandstone-shale sequence that represents withdrawal of the last sea ever to cover this part of the continent.

High cliffs of Tertiary volcanic rocks jut skyward north of milepost 311. They are part of a once-broad lava plateau that covered the Colorado-New Mexico border. We've crossed the Sierra Grande Arch, and now descend into the Raton Basin, where the Great Plains surface, so well preserved farther east, has been cut through and eroded away as streams were strengthened by uplift and by the added rain, snow, and meltwater of Pleistocene time. The Canadian River and its tributaries, issuing from the mountains to the west, carried the rock debris southward and then eastward across Texas, Oklahoma, and Arkansas to the Mississippi, helping to build the Mississippi floodplain and delta.

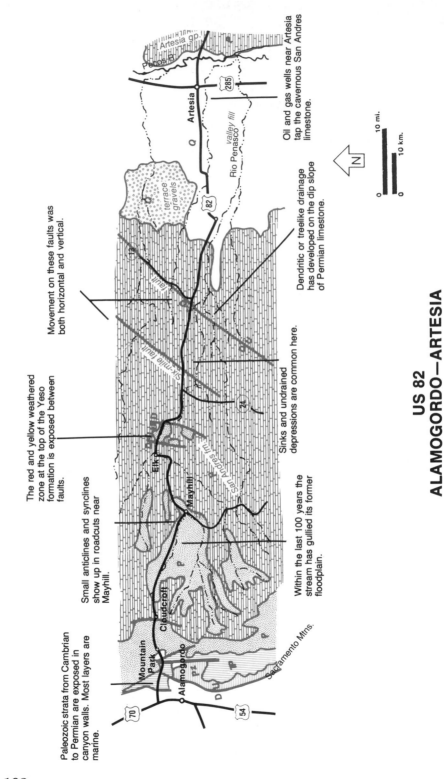

## US 82
## ALAMOGORDO—ARTESIA

Paleozoic strata from Cambrian to Permian are exposed in canyon walls. Most layers are marine.

Small anticlines and synclines show up in roadcuts near Mayhill.

The red and yellow weathered zone at the top of the Yeso formation is exposed between faults.

Movement on these faults was both horizontal and vertical.

Within the last 100 years the stream has gullied its former floodplain.

Sinks and undrained depressions are common here.

Dendritic or treelike drainage has developed on the dip slope of Permian limestone.

Oil and gas wells near Artesia tap the cavernous San Andres limestone.

# US 82
# Alamogordo—Artesia
### 111 mi./178 km.

The Tularosa Basin, between the San Andres Mountains on the west and the Sacramento Mountains on the east, contains more than 6000 feet of valley fill—stream sand and gravel, rockslides, alluvial fans from mountains on either side, and lake deposits rich in salt and gypsum derived from sedimentary rocks of both ranges. The valley has no external drainage, nor has it had in the past. Rain and snow-melt that enter it stay in it. Streams from the mountains sink into porous valley deposits or pool in the deepest part of the valley, there to slowly evaporate, leaving behind their mineral burden: gypsum, salt, and other evaporite minerals.

Paleozoic sedimentary rocks in the surrounding mountains once also contained thick beds of gypsum and salt, showing that they were deposited in a restricted sea and concentrated by evaporation—hence the term "evaporite." With time, much of the gypsum and most of the salt leached out of the rocks and were carried to the valley floor by streams. With nowhere else to go, the minerals accumulated in groundwater of the enclosed basin, becoming more and more concentrated as the groundwater was drawn to the surface and evaporated. The only "sweet" water in the basin today occurs around its margins, where inflow of mountain water pushes back, so to speak, the highly mineralized goundwater in the center of the basin. Even that sweet water contains an unusually large amount of dissolved minerals.

Like the Rio Grande Rift, Tularosa Valley is not river-carved. Caused by sinking between parallel faults, the basin is considered the easternmost part of the Rio Grande Rift. Faults along the base of the Sacramento Mountains form the rift's eastern edge. As we'll see, the east slope of these mountains merges with the Great Plains.

*Hard and soft layers of Mississippian limestone form ledges and slopes in the Sacramento Mountains.*

Paleozoic limestone from which the salt and gypsum come show up well in the Sacramento Mountains. As US 82 leaves US 54/70 to proceed eastward toward the mountains, the stratified nature of the rocks becomes increasingly apparent. The strata stand out on the mountain slopes, emphasized by differential erosion of soft shale and hard limestone layers. Roadcuts expose shaly beds much better than do natural exposures. Between mileposts 5 and 6 a dike cuts through some of the limestone, baking adjacent rocks like reddened pottery.

There is a complete Paleozoic sequence—Cambrian to Permian—in these mountains, but along this highway route Cambrian, Ordovician, and Silurian rocks are covered by alluvial fans, so the lowest rocks visible are Devonian and Mississippian limestone.

Caves of many sizes show up in this limestone. Solution of these caves began in late Mississippian time as the marine limestone was lifted above sea level. We know this because the upper surface of the Mississippian rocks is rough and irregular, with patches of reddish silt and sand, evidence of ancient sinks and development of rugged karst topography similar to that forming today in limestone exposed to warm, humid climates.

194

A turnout just west of the tunnel affords good views of Pennsylvanian strata—more marine limestone, known as the Magdalena group. These strata, as well as Permian rocks above them, match up with those on the San Andres Mountains on the other side of the Tularosa Valley. Here they dip east; there they dip west. Formerly these rocks arched completely across what is now the Tularosa Valley. As the Rio Grande Rift developed, the central part of this arch dropped 5000 feet or more, creating the present valley.

Permian rocks of the great arch formerly included thick deposits of gypsum concentrated as seawater evaporated in the shallows of a restricted seaway. When seawater evaporates, both salt and gypsum are deposited, as well as smaller quantities of other evaporite minerals such as anhydrite and borax. The salt is so soluble that it is washed out by groundwater soon after the rocks are lifted above sea level. The less soluble minerals also gradually disappear. Much of the salt may have returned to the sea during the Mesozoic Era; the less soluble gypsum is still being carried down into the Tularosa Valley, where it has concentrated in and around "Lake" Lucero as the source for the snow white dunes of White Sands National Monument.

Traveling upward through the Permian sedimentary rocks, we see more caves and a number of small springs. The rock is extremely porous, partly because the salt and gypsum it once contained have leached out, partly because the limestone itself is slightly soluble. Permian strata have been divided (from bottom to top, the order in which you will encounter them) into the Abo and Hueco formations—the porous rocks we've been looking at; the Yeso formation—thinly bedded limestone and reddish shale between Mountain Park and Cloudcroft; and the San Andres limestone—a resistant rock that surfaces the long, gentle east slope of the mountains and eventually plunges eastward under the valley of the Pecos River.

The east side of the Sacramento Mountains is much less abrupt than the west side. The gentle dip of the strata can be seen in the walls of Rio Peñasco Canyon, route of US 82. Wherever the canyon widens, small farms have been developed on the soft gray floodplain deposits of the Rio Peñasco. The stream in many places has gullied these deposits to depths of 20-30 feet, probably since cattle-grazing and tree-cutting altered the rate of stream flow and thereby sped up erosion.

Near milepost 34, east-dipping Permian sedimentary rocks are particularly well exposed in a long roadcut. Yellow and orange-brown sandstone and siltstone separate the limestone layers. Small anticlines and synclines show up in roadcuts east of Mayhill.

Cliffs that border the floodplain show interesting erosion patterns: As the stream develops meanders—looplike bends—it carves scalloped indentations in cliffs on either side. Streams cut most rapidly at the outer curves of their meanders—a feature well displayed here.

The limestone layers begin to level out at Chavez City, and from there on east they are increasingly concealed by Quaternary gravel. The bottomlands are deeply gullied here. At milepost 59 the highway climbs out of the Rio Peñasco's little valley, up onto the San Andres limestone surface.

Watch for circular sinks caused by partial collapse of limestone caverns. In some, drainage goes underground through solution channels etched and enlarged along joints in the limestone. Others, their bottoms waterproofed with fine mud, now hold water. Still others guide runoff down into porous underlying strata, where it flows downdip to the valley of the Pecos River.

Just before reaching Hope we encounter Quaternary deposits of the Pecos Valley, sediments brought here by the Pecos River. Small oil and gas wells between Hope and Artesia go through these deposits and into the porous limestone below, which in this area serves as a reservoir for oil and natural gas as well as water.

*In downtown Artesia, the Navajo Refinery processes 30,000 barrels of oil per day, breaking local "crude" oil down into gasoline, kerosene, diesel fuel, lubricating oil, propane and butane gas, asphalt, and chemicals used in manufacturing synthetic fabrics, insecticides, and fertilizers.*

Artesia is at the northern edge of the Permian Basin, a part of Texas and New Mexico that is rich in petroleum resources. As its name indicates, Artesia profits doubly from its geologic setting: The town developed where there is artesian water, which rises in wells without being pumped. The water—from rain and melting snow—enters porous Permian strata on the mountain slopes, as we have seen, and flows down along the sloping strata to the valley, held within certain of the Permian limestones by overlying impervious shaly layers. As it travels down the sloping limestone layers, it develops hydrostatic "head" because of the weight of more water added up-slope—the same process that brings water from a hilltop reservoir down sloping pipes and up into a house. Near Artesia, wells drilled through the impermeable layers provide escape routes for the water as it is impelled upward by this hydrostatic pressure.

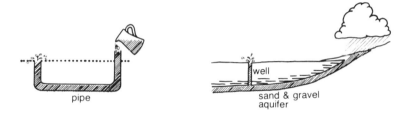

*In an artesian system, the point of entry is higher than the wellhead, so well water rises without pumping. Impervious shale or clay layers prevent the water from rising except where wells are drilled.*

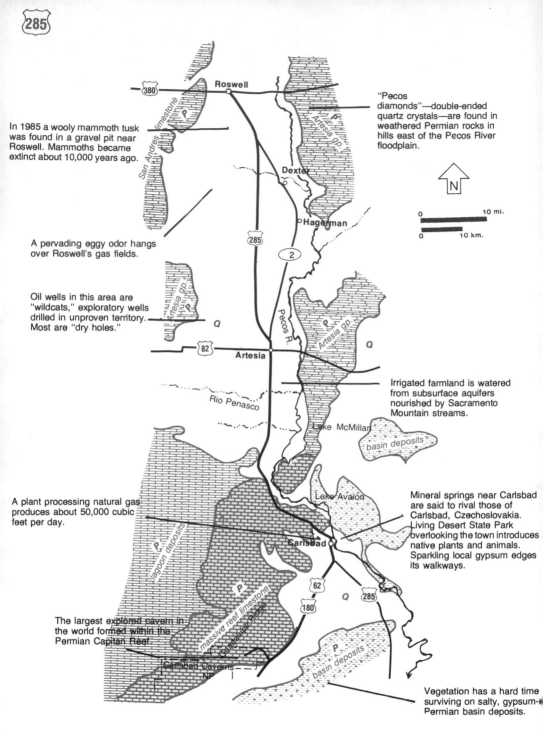

Roswell

380

"Pecos diamonds"—double-ended quartz crystals—are found in weathered Permian rocks in hills east of the Pecos River floodplain.

In 1985 a wooly mammoth tusk was found in a gravel pit near Roswell. Mammoths became extinct about 10,000 years ago.

Dexter

Hagerman

285

2

N

0          10 mi.

0          10 km.

A pervading eggy odor hangs over Roswell's gas fields.

Oil wells in this area are "wildcats," exploratory wells drilled in unproven territory. Most are "dry holes."

82          Artesia

Irrigated farmland is watered from subsurface aquifers nourished by Sacramento Mountain streams.

Rio Penasco

Lake McMillan

basin deposits

Lake Avalon

A plant processing natural gas produces about 50,000 cubic feet per day.

Carlsbad

Mineral springs near Carlsbad are said to rival those of Carlsbad, Czechoslovakia. Living Desert State Park overlooking the town introduces native plants and animals. Sparkling local gypsum edges its walkways.

62

180

285

The largest explored cavern in the world formed within the Permian Capitan Reef.

Carlsbad Caverns NP

basin deposits

Vegetation has a hard time surviving on salty, gypsum-Permian basin deposits.

## US 285
## ROSWELL—CARLSBAD CAVERNS

*Fields of the Pecos Valley are watered from artesian wells like this one south of Roswell.* E.F. Patterson photo courtesy of U.S. Geological Survey.

# US 285, US 62/180
# Roswell—
# Carlsbad Caverns National Park
### 95 mi./153 km. to entry road

Heading south out of Roswell, this highway remains for some time on terraces that border the Pecos River. Lower terraces here, as well as the present floodplain, are farmlands irrigated with Pecos River water. Higher terraces, unirrigated, are grazing land. Note that the river itself is well over on the east edge of its floodplain.

Drilling and pipeline supply companies along the highway, as well as tank cars and trucks on the railroad and highway, announce that we're in "Oil Country, USA." A pervading smell of gas drifts from refineries and gas wells.

Pale gray hills on either side of the river are in Permian limestone, rock that contains enough gypsum to restrict plant growth. The limestone is riddled with solution channels as well, so that water sinks underground readily, further limiting plant growth. The high alkali content of both soil and water leads also to development of caliche, a whitish crusty layer just below the surface.

*Unable to escape upward through impermeable shale, oil and gas collect in upside-down pools in a variety of geologic "traps."*

Oil and gas wells surround the town of Artesia. Those with large "rockers" are oil wells; those with "Christmas trees"—intricate arrangements of pipes, valves, and gauges—are mostly gas wells. Both oil and gas come from porous Permian limestone. In this area, as well as farther south, Permian strata are warped and bent into small anticlines and synclines that make the area a petroleum geologist's paradise. Anticlines act as traps for oil, which collects in porous and permeable rock layers, contained there by impermeable rocks above.

The town of Artesia is named for artesian springs and wells in which water bubbles to the surface without pumping. Water entering porous Permian rocks in the Sacramento Mountains some 60 miles to the west, or draining into the many sinks and small caverns that dot the surface of the Permian limestone between here and the mountains, comes to the surface in springs or is tapped by artesian wells.

As near Roswell, terraces near the river and the modern floodplain are farmed, with artesian wells and the Pecos River providing irrigation water for cotton and other crops.

The highway remains on the river terraces to about milepost 52, then rises onto the limestone surface. There are oil and gas wells here, too. To the east, the edge of the Great Plains shows as a low line of bluffs. Surfaced with Miocene gravels of the Ogallala formation, the Great Plains developed as vast quantities of sand and gravel were carried eastward and southeastward from the Rocky Mountains and the ranges that edge the Rio Grande Rift. Born at the end of Cretaceous time, the Rockies underwent a second uplift in mid-Tertiary time, when they and everything adjacent to them rose an additional 5000 feet or so. Uplift invites erosion, and erosion in one area—the mountains—means deposition in another—the Great Plains. The Plains originally extended right up to the mountains; at that time the Pecos River area was buried under hundreds of feet of sand and gravel. Since Pliocene time, however, the Pecos River has cut through these sediments, carrying much of the rock debris downstream to the Rio Grande.

At Carlsbad we enter an area that geologists call the Delaware Basin, in Permian time a shallow embayment of the sea. Today the one-time embayment is an important petroleum field extending well into Texas.

Mineral springs northwest of the town of Carlsbad gave it its name, as their mineral content, supposedly beneficial to health, rivals that of Carlsbad in western Czechoslovakia.

Carlsbad is also the site of Living Desert State Park, where exhibits include many local rock types. Sparkling white gypsum, deposited as the Permian sea evaporated, is used for edging outdoor gardens. Deep natural ravines in the park expose Permian limestone.

Another resource abundant in Permian rocks near Carlsbad is potash. Valued as fertilizer, potash occurs in thick layers 900 to 1800 feet below the surface, and is mined in highly mechanized underground mines. It was discovered near Carlsbad in 1925 during the drilling of a wildcat exploratory well—one that struck not oil but the first commercial potash in the United States. Like salt and gypsum, potash forms as sea water evaporates. It is used extensively in fertilizers.

The plateaulike highland now to the southwest is Guadalupe Ridge, one of the most remarkable mountain ranges in America. The ridge, continuous with the Guadalupe Mountains farther south, is the work of myriads of small marine plants and animals of Permian time—builders of an ancient reef with many of the characteristics of the Great Barrier Reef of Australia today. Called the Capitan Reef after a prominent peak at the southern end of the mountains, it developed in shallowing water along the edge of the Delaware Basin embayment, where conditions of temperature, salinity, and water movement were just right for rapid growth of reef organisms. Algae, bryozoans, corals, brachiopods, clams and snails, crinoids, and other calcium-secreting plants and animals grew in abundance, building up a ridge of limestone that separated flat-lying sediments of the basin floor from equally flat-lying sediments of a protected lagoon to the west.

Thanks to abundant subsurface information from oil and gas wells, the Capitan Reef is the best-understood fossil reef in the world. As the figures show, the reef is horseshoe-shaped. The gentle curve of Guadalupe Ridge forms one side of the horseshoe. Much of the reef is underground; only a few parts of it, lifted by faulting, are above the present surface, notably Guadalupe Ridge and the Guadalupe and Glass mountains of Texas.

The reef limestone, by its very nature porous, is also intensely fractured as a result of uplift. Water passing through the porous rock, in particular along approximately east-west fractures, has gradually dissolved passages through the limestone. Among these is Carlsbad Cavern, the largest explored cavern in the world, definitely worth a visit.

Rocks close to the highway both north and south of the park entrance road are poorly consolidated, monotonous gray siltstone and limestone—once muddy sediments on the floor of the Permian bay. Interbedded originally with salt, gypsum, and potash, they have bent and broken and collapsed as these soluble substances—particularly the salt—were leached out by rainwater and groundwater. Because they are still highly mineralized, they support little vegetation, enhancing the desert aspects of this part of New Mexico.

# US 380
# Texas—Roswell
### 82 mi./132 km.

The vast plains of eastern New Mexico and panhandle Texas, surfaced with gravel and sand of the Ogallala formation, were established in Miocene and Pliocene time as stream gravels were carried eastward from the Southern Rocky Mountains and new-formed fault block mountains east of the Rio Grande Rift. Once continuous all the way to the mountains, they are eroded from the eastern slopes of the mountains and from the broad valley of the Pecos River. In New Mexico only localized patches of Ogallala formation remain along the eastern border of the state, on the broad plateau of the Llano Estacado.

Between the state line and Roswell the highway goes through northern fringes of one of New Mexico's two major oil-producing provinces. (The other is at the opposite corner of the state, in the San Juan Basin.) Formed from uncounted billions of tiny marine organisms whose bodies were trapped in marine sediments, oil takes millions of years and some very special environmental conditions to develop. Once formed, it tends to migrate from its source rock into and through permeable reservoir rock layers such as sandstone and cavernous limestone. There it collects in various kinds of geologic traps, held by impermeable caprocks such as shale and mudstone.

It is up to geologists to find the trapped oil, and finding it is a complicated business requiring surface mapping, core drilling, subsurface mapping by means of drill cuttings, and study of tiny microfossils brought up in cores and cuttings. Geophysical tools such as electrical logging of wells, study of manmade seismic waves transmitted by rocks below the surface, and studies of local variations in gravity and earth magnetism also help.

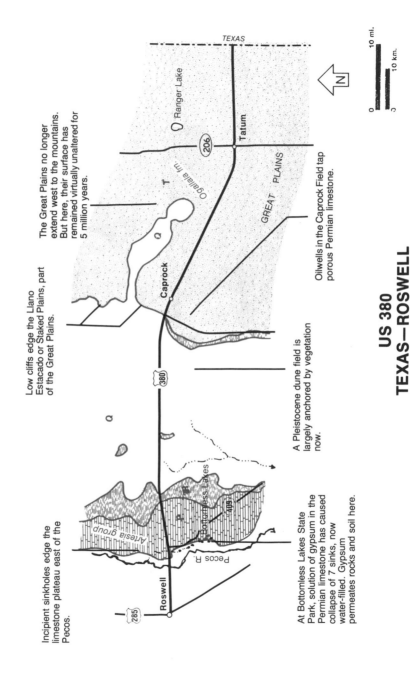

## US 380
## TEXAS—ROSWELL

The Great Plains no longer extend west to the mountains. But here, their surface has remained virtually unaltered for 5 million years.

Low cliffs edge the Llano Estacado or Staked Plains, part of the Great Plains.

Oilwells in the Caprock Field tap porous Permian limestone.

A Pleistocene dune field is largely anchored by vegetation now.

Incipient sinkholes edge the limestone plateau east of the Pecos.

At Bottomless Lakes State Park, solution of gypsum in the Permian limestone has caused collapse of 7 sinks, now water-filled. Gypsum permeates rocks and soil here.

TEXAS

Ranger Lake

Tatum

206

Ogalala fm.

T

Q

Caprock

GREAT PLAINS

N

10 mi.
10 km.

380

Q

Artesia group

Bottomless Lakes

409

Pecos R.

Roswell

285

*Southeastern New Mexico is oil country. Pumps such as this one, rocking gently, bring up oil with every stroke.*

Some of the wells along this highway produce only oil, others only gas. You can frequently tell which wells produce what, because oil is stored in silo-shaped tanks, while gas, stored under pressure, requires strong round-ended silvered tanks. The big "rockers" on many of the wells show that oil is being pumped. Notice the shape of the heads of these rockers—curved so that as they rock, the cable into the well remains centered in the wellhole. In many of the wells, underground gas pressure is great enough to bring oil or gas to the surface, and only small "Christmas trees" of pipes and valves are necessary at the wellhead. Subsurface water may also force oil upward. Both water and gas may be reinjected artificially to maintain pressure. Where pressure is low and oil moves too slowly for continual pumping, the pumps are turned off periodically to allow oil and pressure to reaccumulate—hence the idle rockers.

New holes are being drilled in this area all the time, with portable drilling rigs moved in for the purpose. The tall rigs facilitate lifting out many lengths of drilling pipe each time drill bits—which do wear out—are changed, and addition of more lengths of drilling pipe as a well goes deeper. If a well strikes oil or gas, the drilling rig is moved away and a pump or Christmas tree attached and put to work.

Oil and gas in this part of New Mexico originated in Paleozoic sedimentary rocks, and occur in a number of pools—not pools in the usual sense of the word but underground masses of oil- and gas-saturated rock. Wells vary in depth from about 1000 feet to more than 17,000 feet. Nowadays every effort is made to prevent "gushers," the dangerous and costly fountains of oil common in the early days of the oil industry.

*Ranchers of eastern New Mexico use the state's bountiful winds to raise water for their homes and herds.*

Windmills among the oil wells are another symbol of man's need for geologic knowledge. They tap water in gravel layers of the Ogallala formation. In addition to oil wells and windmills, the Llano Estacado is marked by low circular basins often containing water or mud—the "buffalo wallows" characteristic of parts of the Great Plains. These hollows are shallow sinks caused by solution of gypsum and salt in underlying Permian sedimentary rocks. There are dozens of such sinks between the Texas border and Roswell.

The highway drops down off the surface of the Llano Estacado west of Caprock. The bumpy skyline visible ahead from milepost 196 is a Pleistocene sand dune field—the Mescalero Sands. The dunes are mostly stabilized by vegetation now, though still active sand surfaces can be seen to the south from milepost 194. Prevailing winds in this area are from the west, and the dunes show a classic dune profile—a long, gentle windward slope, and a shorter, steeper leeward one.

Passing through another oil field, where pumps are smaller because reservoir rocks are nearer to the surface, we encounter some dark red Triassic sedimentary rocks and then light tan and pinkish tan Permian rocks belonging to the Artesia formation. The latter rock unit is well exposed along the edge of the Pecos Valley.

Along the valley bluffs, particularly at Bottomless Lakes State Park a few miles south of milepost 164-165, are many sinks caused by

*The edge of the Llano Estacado is clearly marked by a line of bluffs capped with Ogallala Formation gravel cemented with caliche.*

206

*Bottomless Lakes near Roswell occupy sinks formed by collapse of underground caverns, in turn produced by solution of gypsum in Permian rocks.*

solution of gypsum and salt from these Permian rocks, with resulting collapse of overlying parts of the formation. There is lots of gypsum still here—some of it right at the surface. Several of the water-filled sinks lie out on the Pecos River floodplain; others are in the edge of the bluffs themselves. A few of the lakes are bordered by crusts of gypsum and other salts—deceptive crusts not strong enough to bear the weight of humans or automobiles. The lakes are not really bottomless, but they *are* deeper than you would expect—as deep as 90 feet. Rocks between the sinks are highly distorted and broken up, also because of solution of gypsum and salt.

Closer to Roswell, the highway approaches the Pecos River, with lumps and stringers of white gypsum visible in roadcuts. With most of its water already drawn off for irrigation, the Pecos is now a shadow of its former self. Containing lots of gypsum and other salts, the floodplain on either side of the river is not much good for farming. It is used for grazing, however.

Originally a ranching and cattle-shipping center, Roswell is a petroleum town now—complete with the sulfurous fragrance of natural gas. Its water comes not from the Pecos River but from the Capitan, Sierra Blanca, and Sacramento mountains to the west. Flowing through cavernous, gently east-dipping Permian limestone, the water reaches the Pecos Valley with a build-up of hydrostatic pressure, so that it rises in wells without pumping. The discovery of this artesian water in 1891 led to explosive growth of agriculture in the Pecos lowlands.

*Bottomless Lakes near Roswell.*

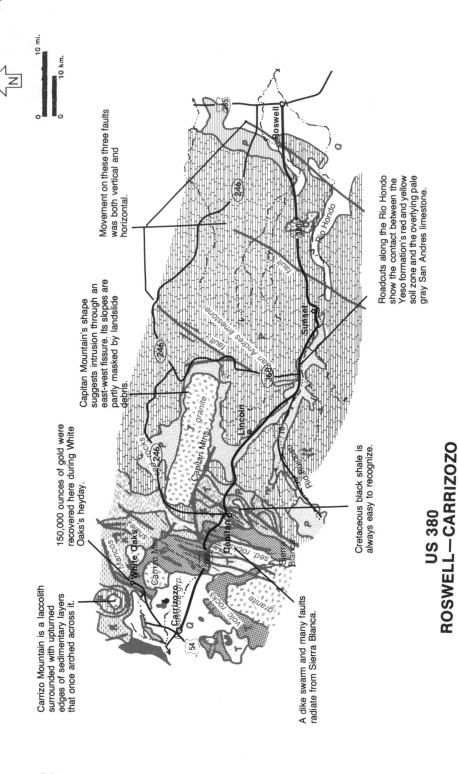

Carrizo Mountain is a laccolith surrounded with upturned edges of sedimentary layers that once arched across it.

150,000 ounces of gold were recovered here during White Oaks's heyday.

Capitan Mountain's shape suggests intrusion through an east-west fissure. Its slopes are partly masked by landslide debris.

Movement on these three faults was both vertical and horizontal.

Roadcuts along the Rio Hondo show the contact between the Yeso formation's red and yellow soil zone and the overlying pale gray San Andres limestone.

A dike swarm and many faults radiate from Sierra Blanca.

Cretaceous black shale is always easy to recognize.

## US 380
## ROSWELL—CARRIZOZO

*A former sink, filled with silt, sand, and broken rock material, cuts limestone layers of a highway cut east of Roswell.*

# US 380
# Roswell—Carrizozo
## 89 mi./143 km.

In the fertile Pecos River Valley, Roswell receives only 13 inches of rainfall a year—not enough to nourish its fields and orchards. Artesian water was discovered here in 1891. Rain falling on the broad limestone slope east of the Sacramento Mountains sinks quickly into underground caverns dissolved in the limestone, to begin a four- to five-year journey eastward to the Pecos Valley. Since it enters these channels at higher elevations, the water is under pressure and flows without pumping from wells that tap the cavernous limestone. Additional irrigation water now comes from the Pecos River.

Climbing out of the valley the highway crosses several poorly defined terrace levels, with fields nourished by this water. The final rise is to a stony, unfarmed limestone surface thinly covered with gravel.

Exposed in small roadcuts is the San Andres limestone, whitish or yellowish Permian marine rock, the aquifer for Roswell's water. Originally deposited in flat-lying layers on the sea floor, it now tilts eastward; near mileposts 328, 310, and 294 it arches into narrow anticlines. Evidence of underground passages that honeycomb the San Andres limestone can be seen from the highway: sinks formed as cavern roofs collapsed. The formation contains many fossil shellfish.

The cone-shaped end of Capitan Mountain dominates the skyline northwest of the rest stop at milepost 315. This mountain's shape changes as it is viewed from different points along the highway. It is a laccolith, an intrusion that squeezed between sedimentary layers, doming up those above it. Farther west is the towering summit of Sierra Blanca, an eroded composite volcano whose violent eruptions about 35 million years ago threw out large quantities of broken volcanic rock and volcanic ash.

*Crested with snow, Carrizo Mountain is a laccolith in which magma squeezed up along a long, narrow fissure.*

Much of the ground surface along the highway west of the rest stop is caliche. Used for road material here, caliche forms a good aggregate that tends to harden with time.

At milepost 303 the highway drops into the valley of the Rio Hondo, with good exposures of San Andres limestone and, under it, the Yeso formation, also Permian. Individual limestone layers in the San Andres limestone alternate with thin beds of shale. Valley walls reveal many small caves.

Mostly pink sandstone and siltstone, the Yeso formation contains some limestone and layers of gypsum. It was deposited in an inland sea where high evaporation caused precipitation of gypsum and salt. Its contact with the San Andres is marked by the yellowish-weathering sandstone visible west of milepost 297 and near the Rio Hondo bridge. Being very soluble, the salt has leached away; much of the gypsum is probably gone, too. The remaining gypsum is weak and soft, and the rest of the formation is crumpled and commonly concealed by slumps.

The Rio Hondo's floodplain, with small farms and orchards, is deeply gullied by the present river. Gullying started during the

1860s, possibly as a result of overgrazing. Gullying provides channels for runoff and lowers the water table of the surrounding region, so that ranchers of today suffer from the sins of their forebears. Low terraces edge the valley.

West of Hondo Junction, US 380 follows the valley of Rio Bonito, and we see more exposures of the Yeso and San Andres formations. The San Andres limestone is essentially flat-lying, but the Yeso is thoroughly crumpled into the so-called Lincoln folds, which are exposed for about seven miles in bluffs north of the river. Here, too, the crumpling was caused by solution of gypsum and salt, perhaps augmented by intrusion of the Capitan Mountain laccolith or slow downhill sliding of the San Andres limestone over Yeso strata.

Side streams cut deeply into the soft Yeso formation. Most of them flow only during and after heavy rains. When they disgorge into the valley they deposit bouldery alluvial fans that scallop the valley margins. Where the highway crosses these fans their characteristic medley of poorly-sorted pebbles, cobbles, and sand can be seen close up.

East and west of the historic town of Lincoln, big blocks of San Andres limestone have tumbled down across the Yeso formation. Both formations rise westward to the summit of an anticline at milepost 92. There they bend over, and the Yeso formation and its yellowish sandstone disappear underground. In this area, some of the many caves in the San Andres limestone have yielded prehistoric Indian artifacts.

Slope-forming reddish siltstone, limestone, and gypsum exposed west of the anticline are part of the Artesia formation. By milepost 89, near the Fort Stanton junction, these Permian rock units, too, disappear underground. Their place is taken by the Triassic Santa Rosa

*Gullies such as this one have been excavated within the last 100 years, probably because grazing removed moisture-conserving vegetation and lowered the water table.*

211

sandstone, with a pebbly conglomerate layer about six feet thick as its base. Many formations have a basal conglomerate, evidence that boundaries between formations may represent times of erosion. The conglomerate contains pebbles of chert, quartzite, and igneous rocks not known in this region—smoothly rounded pebbles which must have been carried many miles by Triassic streams and rivers. The rest of the Santa Rosa sandstone is a beach deposit of clean quartz sand with the uniform grain size typical of beach sand.

From the pass west of Fort Stanton junction, Capitan Mountain shows up well. Tertiary intrusive rock that forms it pushed its way between the San Andres limestone and underlying rock layers, doming up the limestone and younger rock units that lie above it. Most of the domed-up rocks are now eroded away, but their cut edges appear around the mountain base wherever they are not covered with landslide and rockfall debris. Most laccoliths are approximately circular, but this one is 13 miles long and four miles wide—a shape that suggests that its magma rose along an east-west fracture. Intrusion of the magma altered adjacent sedimentary rocks, enriching them with metallic vapors, creating ores of iron and other minerals.

As the highway continues westward younger sedimentary rocks are exposed: redbeds of the Triassic Chinle formation and above them cliff-forming Upper Cretaceous Dakota sandstone. Jurassic and Lower Cretaceous rocks are absent and may never have been deposited here.

The Chinle formation, a slope-former, is a floodplain deposit that stretches from New Mexico into Arizona and Colorado. Its colorful mudstones, some of them rich in decomposed volcanic ash, contain fossilized wood and skeletons of amphibians and reptiles. The Dakota sandstone, a cliff-former, is a near-shore and beach sandstone that reflects gradual encroachment of the sea. Though the Dakota sandstone is light tan in color, exposed surfaces are usually dark with lichens and desert varnish.

Farther west we come on dark gray Mancos shale. At the contact of the two formations layers of sandstone alternate with layers of shale. The Mancos shale contains upper Cretaceous fossil clams and ammonites, shelled relatives of today's octopus and squid.

West of milepost 86 watch for dikes. There are some 1000 dikes in this area, and hundreds of them cross the highway in the next 12 miles, many forming distinct ridges.

Capitan was originally a coal-mining and iron-mining center. The largest iron deposit is about six miles north of town; coal mines are just west of town, in rocks of the Mesaverde group—sandstone, siltstone, and coal deposited along the shore of the retreating Cretaceous sea. Parts of this formation contain shells of oysters and large clams.

West of Capitan's coal mines we encounter sugary white Tertiary sandstone mixed with dark red and purple siltstone and thin conglomerate layers. The variety of rock types suggests a varied environment; the rock was deposited on river floodplains, in short-lived lakes, and in landlocked bays and lagoons.

At Indian Divide the highway crosses an almost north-south fault that brings rocks of the Mesaverde group to the surface again. To the west lies the Tularosa Valley, an undrained basin that is part of the Rio Grande Rift. Visible far out in the valley is the black swath of the Valley of Fire lava flows.

Small peaks north of the highway are some more Tertiary intrusions. There is a huge alluvial fan at the base of these peaks. Gold and silver were mined in these hills in the 1900s; with gold's rising prices a few years ago, interest in the area was rekindled.

As the highway descends there are roadcut exposures of Mesaverde sandstone and coal, again followed by purple and white Tertiary sandstone. Both are cut by numerous dikes. A particularly large dike crosses the highway near milepost 75 and takes off cross-country as a prominent ridge. Farther west, the highway descends a broad alluvial apron to Carrizozo.

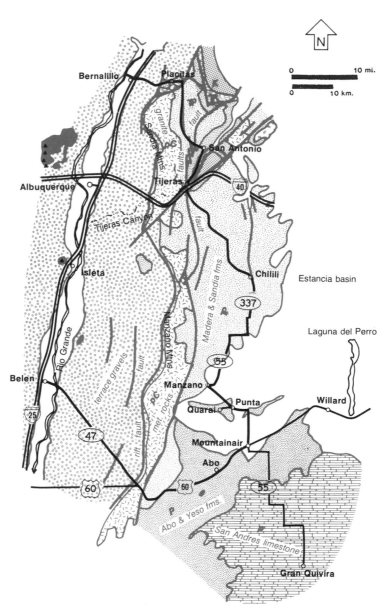

## SANDIA MOUNTAINS LOOP

# Sandia Mountains Loop
## 119 mi./191 km.

This trip circles the Manzano and
Sandia mountains in a counterclockwise direction,
starting at the I-25/NM 47 junction near Belen.

Leaving Interstate 25, NM 47 crosses the fertile, well watered Rio
Grande floodplain. Valley fill here is on the order of two miles thick,
and terraces on either side of the valley show that it has been thicker
in the past. The youngest 90-100 feet of sediment, forming the present floodplain, have been deposited within the last 10,000 years.

The highway rises over several broad alluvial terraces, each representing a period of relative stability in the history of the river, when
downcutting ceased for a time and the river deposited enough sand
and gravel to form a floodplain. Farther north, in Colorado and
Wyoming, similar terraces tie in with glacial deposits; here they may
reflect the great deluges and heavy runoff that corresponded to
northern glacial advances. The highest terrace is the oldest, with
successively lower ones successively younger. Terrace surfaces are
almost level, though the highest shows some of the hummocky topography of former sand dunes. Gravel that forms the terraces is derived
mostly from upstream sources, brought in by the Rio Grande, though
the uppermost terrace also includes pebbles from the nearby Manzano Mountains. The uppermost terrace blends indiscernibly with a
surface cut into the rock mass of the Manzano Mountains
themselves—a true mountain pediment.

215

Near the southern end of the Manzano Mountains, we join US 60 to cross the major fault zone—hidden beneath alluvial fans—that marks the edge of the Rio Grande Rift. We then climb to the divide between the Manzanos and Los Pinos mountains farther south. The west faces of the ranges are made up of Precambrian rock, mostly gneiss and schist, whereas the eastern slopes consist of Pennsylvanian sedimentary rock. In the Manzano Mountains Precambrian rock is pushed up along a reverse fault so that it is partly *over* the Pennsylvanian rocks. Established in late Cretaceous time, the reverse fault is older than faults that border and shape the Rio Grande Rift.

*The broad, gently sloping bajada west of the Manzano Mountains is partly stream-deposited, partly eroded into the mountain base.*

There are no Cambrian, Ordovician, Silurian, Devonian, or Mississippian strata here; any that existed were removed by erosion before Pennsylvanian rocks were deposited. Just in these two ranges we can deduce at least four periods of uplift or mountain-building:

1) at the end of Precambrian time, before any Paleozoic sedimentary rocks were deposited;

2) before Pennsylvanian deposition, with removal of Cambrian to Mississippian rocks from an uplifted area;

3) in late Cretaceous time when reverse faulting occurred in the Manzano Mountains area; and

4) in Tertiary time as the Rio Grande Rift subsided between its bordering ranges.

In addition, there were apparently many episodes of faulting, folding, and general squeezing in Precambrian time, a good deal harder to unravel. There are exposures along this highway of several types of Precambrian rock, all seemingly separated by faults and unconformities: schist formed by metamorphism of fine-grained shale and mudstone, greenstone formed by metamorphism of volcanic rocks, and quartzite formed by metamorphism of sandstone.

East of the pass, US 60 descends the slope of a large alluvial fan from Abo Canyon, deeply channeled near the mountains by the well fed, fast-moving mountain stream. This fan coalesces with its neighbors to form an alluvial apron, or bajada, all along the east side of the mountain.

Near milepost 185 we cross the reverse fault that separates Precambrian rocks from the Pennsylvanian Madera formation, a unit made up of many thin layers of fine, gray marine sandstone, mudstone, and limestone.

Near milepost 186 Permian sedimentary rocks appear—brown and dark red mudstone and sandstone of the Abo formation, deposited on a river floodplain or a delta near the Permian sea. Sandstone layers in this formation show types of crossbedding characteristic of channels and bars of a meandering river. Some of the layers are nicely marked with ripple marks; others display mudcracks formed by shrinkage as the sediments dried. Copper and a little uranium have been mined near milepost 192 from some of the channel sandstone.

The ruins of Abo Pueblo, half a mile north of the highway, show that pre-Spanish as well as Spanish builders appreciated the way that sandstone layers of the Abo formation break naturally along joints, giving neat rectangular blocks suitable for masonry. The Abo ruins, with Gran Quivira and Quarai, make up Salinas National Monument. The Abo formation dips distinctly eastward here, its more resistant beds forming cuestas west of Abo and extending north and south along the base of the mountains.

Just east of the ruins is a younger, more widespread Permian unit, the Yeso formation, with pinkish, crossbedded sandstone, thin-bedded brown mudstone, and abundant gypsum.

East of the Manzano and Sandia mountains is the Estancia Basin, lake-filled in Pleistocene time, wth marks of its old shorelines still visible on aerial photographs. To the south is Chupadera Mesa, capped with Permian limestone of the San Andres formation, a Permian unit above the Yeso formation.

Between Mountainair and Willard the highway crosses the silty floor of Pleistocene Lake Estancia, a fertile bottomland tilled by the agricultural people who built Abo, Quarai, and Gran Quivira. Sand hills around the margins of the valley are the remains of old dunes. Fine, fertile soil of the old lake floor, once able to support the three Salinas villages, was virtually destroyed by 19th- and 20th-century deep plowing. Problems were augmented by droughts of the 1930s and 1940s. The basin now continues to deepen as wind erodes what is left of the lake sediments.

217

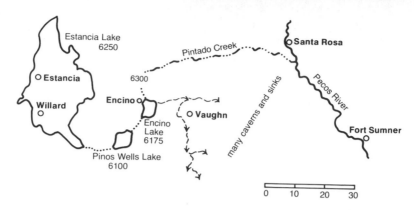

*Present elevations show that Pleistocene Lake Estancia could not have drained eastward via Pintado Creek. Its waters may have escaped through limestone caverns in the area southeast of Encino Lake.*

Along the east side of Estancia Basin are several interesting salt lakes, among them Laguna del Perro. They were well known to pre-Spanish Indians of the area, who found salt from the lakes a valuable commodity for their own use and for trade. With the coming of the Spanish, salt mined from these lakes was shipped to Mexico for flux in silver smelters. The ponds lie between sand dune ridges.

North of Mountainair the highway skirts the mountains, crossing coalesced alluvial fans and offering good views of east-dipping Pennsylvanian rocks on the mountain flank. Thick crusts of caliche on the alluvial fans suggest that they are quite old as fans go— probably Pleistocene.

Big roadcuts near milepost 108 reveal nearly flat-lying Pennsylvanian rocks, with some minor very gentle warping.

Between mileposts 114 and 115, a river cut shows a strongly marked anticline, in places offset by faults. The black shale below the sandstone cliffs, near the junction with Interstate 40, is part of the Pennsylvanian Madera group.

North of Interstate 40 the structure of the back of the Sandias can be seen—more steeply tilted east-dipping Pennsylvanian rocks forming a relatively smooth, forested surface. This stretch of road has been christened the "Turquoise Trail" because it passes through the Cerillos Hills, where turquoise has been mined since prehistoric time. Used as a gem and ornamental stone for more than 8000 years, turquoise forms as rounded or bumpy nodules or thin crusts deposited by underground water circulating through certain kinds of volcanic rock. Oddly, its occurrence is limited to arid regions. Sky-blue varieties such as that found in the Cerrillos Hills are the most valuable; their color comes from minute amounts of copper compounds.

218

Golden sprang up during a short-lived 1879 gold rush. Production of gold was hampered by lack of water, but silver-lead ores were mined for a time, even sent to England for smelting. Coal mined near Madrid was used for small-scale local smelting; it also supplied the growing city of Santa Fe. Dumps of the old underground mine can be seen near the town, as can a few old miners' cabins. The coal here was anthracite, hard coal, quite rare in the West. It occurred in a thin seam that could not be mined economically today.

North of Cerillos the road travels among some vertical sandstone beds interspersed with red mudstone layers. Several small peaks of volcanic rock, as well as a dike about half a mile west of the road, show that the volcanic area of the Ortiz Mountains and Cerillos Hills extends north here.

The high-level surface around the Ortiz Mountains is a partly eroded, partly stream-deposited surface about 400 feet higher than the Rio Grande. Geologists recognize this Ortiz Surface in a number of places in northern New Mexico.

*Complete this loop drive using logs for Interstate 25 from Santa Fe to Albuquerque and, in part, Albuquerque to Socorro.*

# V
# National Parks and Monuments

The national parks and monuments in New Mexico offer unusual opportunities to look at rocks in a more leisurely fashion. Though the focus of some of them—Bandelier, Gila Cliff Dwellings, El Morro, and Salinas National Monuments—is in their archeologic records, all those described here display interesting geologic features, too.

In all parks and monuments, rocks, fossils, minerals, and animals are protected; no collecting is permitted.

## Bandelier National Monument

Prehistoric ruins at Bandelier occupy a canyon cut by the Frijoles River in volcanic tuff, once the ashy outpouring of explosive volcanic eruptions that almost completely demolished the large volcano from which they came.

Only the base of the Jemez volcano, which existed here about a million years ago, now remains—the rim of the Valles Caldera at the center of the Jemez Mountains. Like many other New Mexico volcanoes, the Jemez volcano rose along faults that edge the west side of the Rio Grande Rift. With alternate layers of thick lava and ash, results of alternating fairly quiet and quite explosive eruptions, the mountain was a composite volcano, probably of the shape and size of Mt. St. Helens before its 1980 eruption. It rose above a base of older volcanic rock, some of which can be seen in the lower gorge of the Frijoles River. A little more than a million years ago it reached its maximum height.

*The welded tuff of Frijoles Canyon erodes into strange forms, its crusted surface penetrated by numerous hollow alcoves.*

Shortly after that, about one million years ago, the volcano burst forth with two extremely violent eruptions, spewing out incredible volumes of volcanic gases, ash, pumice, and broken rock. The Mt. St. Helens 1980 eruption was child's play by comparison: The Jemez eruptions released more than 50 cubic miles of rock material— roughly 100 times that discharged by Mt. St. Helens!

Ash clouds drifted as far as Iowa, Oklahoma, and Texas. Other ash sped down the flanks of the volcano in incandescent avalanches that finally came to rest far down on the mountain slopes. Still very hot, the incandescent particles welded together, forming a firm yet porous rock. The ash flows of the second great eruption now form the Bandelier tuff that walls Frijoles Canyon.

The explosions to some extent depleted the magma chamber far beneath the volcano. No longer supported from below, the mountain, ringed by fractures, collapsed. Its subsidence produced, at the surface, a vast caldera, an almost circular, cliff-ringed pit 14 miles across—the Valles Caldera, not within the national monument but well worth a visit.

Much of the Bandelier tuff represents two thick layers of volcanic ash. Their upper and lower margins, where they cooled most rapidly,

*Two layers of light-colored, welded ashflow tuff and one layer of darker lava form cliffs edging the Pajarita Plateau. Slopes develop on less well consolidated tuff.*

remained fairly soft and easily eroded; their centers, cooling slowly over many decades, became hard and resistant. Most of the soft upper part of the tuff has eroded away completely; the hard center of the upper ash flow forms the mesa surface around Frijoles Canyon.

Many features of the tuff can be seen near the ruins and along the trail downstream toward the Rio Grande. The rock is composed of bubbly fragments of volcanic ash, hard shards of volcanic glass, and irregular lumps of other volcanic material. In places it fills channels cut in older lava flows. Elsewhere it is studded with volcanic bombs rounded as they spun through the air.

On the whole, the tuff is weather-resistant, with a hard outer shell that developed as moisture sank into the rock, absorbed silica minerals from the tuff, and then was drawn to the surface and evaporated by warm sunshine and dry air. Scattered through the tuff are natural alcoves hollowed out by rain, wind, and snow; some of those near the ruins were enlarged by the pueblo-dwellers of Frijoles Canyon. In places weathering has followed vertical joints to the extent that conical rock forms locally known as "tent rocks" have developed. Tent rocks may also form by weathering where volcanic fumes rise through porous ash, cementing some of it more tightly.

Downstream from the ruins the canyon walls are composed of dark lava flows that predate the Bandelier tuff. The stream, struggling with this harder rock, has cut a narrower defile. Red-baked soil zones can be seen beneath individual flows, and little green olivine crystals

222

appear in the basalt. Olivine basalt, known to have come from the Earth's mantle deep below the solid crust, tells us that the faults along which it rose—faults that edge the Rio Grande Rift—reach all the way through the crust.

## Capulin Mountain National Monument

Capulin Mountain, a classic cone-shaped cinder cone, is one of many interesting volcanic features of northern New Mexico. A road leads up the cone to the margin of its crater; a trail circles the crater rim, affording good views of surrounding features. At the Visitor Center an introductory film includes movies of the 1943 eruption of Paricutin, a cinder cone in Mexico that in a few months developed from a crack in a cornfield to a mountain comparable to Capulin.

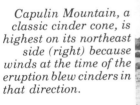

*Capulin Mountain, a classic cinder cone, is highest on its northeast side (right) because winds at the time of the eruption blew cinders in that direction.*

The slopes of Capulin Mountain are composed of bubbly pellets of volcanic material thrown from the central crater. Near the crater, some of the partly molten pellets, or cinders, fused, forming a welded rock, but elsewhere they lie loose on the slopes. Larger fragments— volcanic bombs spun into football shapes—bounced down the cinder slopes. Because prevailing winds blew from the southwest at the time of the eruption, as they do today, the crater rim is highest on its northeast side.

Several of the lava flows that surround Capulin came from the west base of the mountain itself. Each flow, as it cooled at the edges, built ramparts of hardened lava, which then guided the course of the flow.

223

*Sloping layers of volcanic cinders show clearly along the road to Capulin Mountain's summit crater.*

Some of the ramparts surround the picnic area. Most of the flows are short and blocky, formed of thick, sticky lava. However one of the youngest flows, curving around the base of Capulin Mountain, shows surface features of quite fluid lava, and strongly resembles hardened black taffy.

The cinder cone-lava flow combination is common. Lava rich in gas from groundwater bubbling into it creates a foam that floats on less gassy lava in the underground magma chamber and the conduit of the young volcano—the same type of separation and layering seen in a foaming bottle of beer or soda. When the bottle is uncorked, so to speak, foam rushes up the escape route and shoots skyward. As globs of foam cool they fall back around the vent, building up a cinder cone. Then, as the gas is used up, less gassy magma rises through the conduit, eventually escaping as lava flows. Depending on the strength of the cinder mountain as well as on the volcanic forces that push the lava upward, flows may rise into the main crater or may escape via cracks and fissures from the base of the cone, as at both Capulin and Paricutin.

*Capulin's crater has partly filled with sliding pumice from its walls. Coarse blocks surround the original vent.*

*A small lava dome protrudes above a grass-covered lava flow northwest of the cinder cone.*

From the summit of Capulin Mountain several other small cinder cones are in view. Sierra Grande, off to the southeast, is a good example of a shield volcano built almost entirely of moderately thick lava. Lava-capped mesas dot the surrounding area. Some are long and narrow and result from fluid flows that coursed down pre-established stream valleys. The lava displaced the streams, which moved over to erode softer rock on either side of the former valley. Eventually this softer rock wore away, leaving only the sinuous lines of the lava flows to mark the position of the earlier streams.

Volcanic features of this area lie on top of the Pliocene surface of the Great Plains. The oldest volcanic rocks are those of Mesa de Maya along the Colorado-New Mexico border, visible on the skyline to the north. Once far more extensive, they formerly covered the Capulin Mountain area. Next oldest are the Clayton basalts, which occur under and around Capulin Mountain and extend east almost to the Oklahoma line.

Capulin Mountain itself erupted between 8000 and 2500 years ago, at a time when Folsom Man, named for an archeological site near the little town of Folsom a few miles north of Capulin, was already on the scene, hunting now-extinct bison, camels, musk oxen, and giant sloths with carefully shaped stone spear points. Folsom Man may well have witnessed the eruption of Capulin. The charcoal of his fires, analyzed by the carbon-14 method, has helped to date Capulin's cinders and volcanic ash.

Capulin's rim offers good views of the surrounding country, including the saw-toothed, usually snow-capped Sangre de Cristo Range to the west—the southern tip of the Rockies, lifted along the faults that edge the Rio Grande Rift.

# Carlsbad Caverns National Park

Celebrated for the size of its underground rooms and corridors as well as for the beauty of its stone ornaments, Carlsbad Cavern was formed by a two-cycle process that 1) excavated the cave itself, and 2) decorated it with stalagmites, stalactites, and other dripstone and flowstone features. Paved trails, stairways, and radio-guided tours enable visitors to see the cave at their own pace. Undeveloped New Cave can be visited on guided tours.

*Visitors admire the decorative stalactites and stalagmites of Carlsbad Cavern.*
W.T. Lee photo courtesy of U.S. Geological Survey.

Caverns such as these develop by gradual solution of limestone. Rain and snow, absorbing carbon dioxide from air and soil, become slightly acid—an age-old process that long predates man's pollution of the atmosphere. As such slightly acid water seeps into limestone, it little by little dissolves away the calcium carbonate of the limestone. Following joints, fissures, and the layering of the rock, the water very gradually enlarges its channelways. Greatest solution occurs at and just below the water table, the surface below which the rock is completely saturated.

If the water table drops, leaving the enlarged channelways dry, calcium-charged water seeping in from above tends to deposit its mineral load as travertine, a form of calcite. Where water drips from the ceiling, stalactites form, and below them, catching the drip, stalagmites rise from the floor of the cave. Water flowing down cave walls or sheeting from ceilings builds smooth flowstone surfaces and delicate draperies. Where pools form, they become edged with travertine lilypads.

One of the interesting features of these caverns, and one which no doubt contributed to the great size of the caves, is that they developed not in ordinary sea-deposited flat layers of limestone, but in the massive limestone of an ancient reef. In Permian time, some 290-240 million years ago, a shallow tropical sea covered much of the southern part of the continent. Here in what is now New Mexico there were three bays at the margin of this sea. And encircling the middle bay, where conditions of depth, temperature, light, and wave action were just right, there was a population explosion of marine plants and animals, particularly of calcareous algae. Growing in colonies, the algae cemented themselves together, incorporating coral and other animal shells, until a solid reef was formed. With passing centuries, the reef grew upward and outward, merging with basin deposits to the east, and became partly buried by lagoon deposits to the west.

The entrance to Carlsbad Cavern is surrounded by flat-lying layers of limestone deposited in the Permian lagoon. The steep slope along the east side of Guadalupe Ridge is the seaward side of the great reef—prominent now partly because of faulting and uplift, partly because erosion has removed softer deposits of the ancient bay.

Reef growth eventually stopped, probably because the nearly landlocked embayment became too salty for reef organisms. When, many millions of years later, the reef was lifted high above the sea, the inflexible masses of reef limestone were cracked and broken, and the stage was set for development of Carlsbad and other caverns.

*The cavern entrance, created by collapse of part of the cavern roof, is framed with lagoon limestone deposited behind Capitan Reef.* National Park Service photo.

Uplift of Guadalupe Ridge and the Guadalupe Mountains to the south, also part of the reef, occurred in several installments starting about 12 million years ago. As the great reef was lifted, solution began—solution that tended to concentrate at the level of the water table, and that opened up the highest of the cavern levels. Then, with each further upward movement, as the water table occupied lower and lower positions, lower channelways were dissolved. Every prominent level in the cavern represents a period of time when the water table remained fairly constant. Immersion was a product of interplay between uplift and climate, and at times once-dry portions of the cavern were resubmerged.

The floorplan of the cavern was determined by joints in the massive limestone. Most of them are vertical, and they are arranged in parallel sets that trend either northeast—parallel to the reef escarpment—or southeast. As we might expect, these are the directions of most of the passageways and rooms within both Carlsbad Cavern and other park caverns.

As water drained from any one level, ceilings and partitions must certainly have collapsed, enlarging some parts of the caverns, reducing others. Then, fairly late in the history of the caverns, second-stage processes began. Calcium carbonate-rich water seeping down from above—some in tiny droplets, some in thin sheets—began to deposit its dissolved minerals on cave surfaces. Very slowly, stalactites formed. Water dripping from their tips in turn built up stalagmites on the floors of the caverns. Through thousands of years, inch by inch, draperies and intricate stone waterfalls developed, along with many

*A plan of the cavern shows that solution tended to follow two sets of joints.*

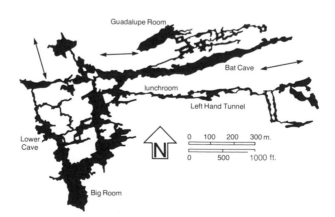

*Stalactites and stalagmites, growing by tiny additions of calcium carbonate, eventually join to form columns.* National Park Service photo.

other types of ornaments. Cave popcorn formed when the water level rose and rooms and corridors were reflooded. Then, as the climate became drier, the water table dropped again, leaving the caverns as we see them today.

*Curly helictites seem to defy gravity. Capillary action and fluid pressure, working through tiny center openings, control the direction of their growth.* National Park Service photo.

Recently discovered evidence points to other methods of cave development that may have played a role in creating caverns in this area. The abundance of gypsum in Lechuguilla Cave in particular (open only to experienced cavers contributing to the knowledge of the cavern), and in the Guadalupe Mountains in general, suggests that sulfuric acid, derived from pyrite in lagoon limestone, may have played an important role. The sulfuric acid process would convert limestone (calcium carbonate) to gypsum (calcium sulfate), which then would slowly decompose, hollowing out caverns without the necessity of their being submerged below the water table. This method of cave formation would also explain some large blocks of gypsum and gypsum ornaments within Carlsbad Cavern itself.

*The rain-fed pool at El Morro National Monument attracted Indian, Spanish, and Anglo travelers with its ever-reliable supply of water.*

## El Morro National Monument

Sited where Triassic and Jurassic sedimentary rocks slope southward off the Zuni Mountains, this national monument presents a vivid picture of early Spanish and American travelers, their names and dates inscribed on the jutting bluff that gives El Morro its name. The cliffs are composed of cream-colored Zuni sandstone. Broad, sweeping crossbedding and the even size of the sand grains, as well as their rounded, finely frosted surfaces, tell us that this rock is a

230

*The dark stain of lichens, located along a past or present line of seepage, covers some Indian petroglyphs. A Spanish inscription shows faintly at the upper right.*

product of sand dunes, part of a Jurassic desert in many ways similar to the modern Sahara.

This massive rock overlies easily eroded Triassic mudstone that forms a ringlike "racetrack" valley around the Zuni uplift, and that as it erodes, tends to undermine the Zuni sandstone. The presence of water, as for example in the deep pool below the inscriptions, further weakens the cliff base and of course provided the main attraction for early travelers following the thirsty trail across this part of New Mexico. Even before 1605, the date of the earliest Spanish inscription, the waterhole served two mesa-top pueblos, now in ruins.

Exposed surfaces on the Zuni sandstone show many interesting weathering characteristics. The smoothest, lightest surfaces are the most recently exposed. Several large archlike overhangs mark places where rock has fallen away in a pattern often seen in thick deposits of dune-formed sandstone. Elsewhere in the Southwest, some similar arches are free-standing, as in Rainbow Bridge and Arches national monuments in Utah. Vertical streaks on the face of the cliff mark lines where water at times trickles from above, encouraging growth of dark lichens and mosses. Where water seeps through rather than over the rock, emerging at some of the many horizontal planes, small caves develop. Vertical joints widen as water seeps down through them, dissolving the small amount of cementing material that holds the sand grains together.

Capping the cliff is the oldest Cretaceous sedimentary layer, the Dakota sandstone, a unit deposited along the western shore of an advancing Cretaceous sea. The trail up and over the mesa top will give you a chance to see this shore sandstone close up, and the contact between it and the Zuni sandstone. A guidebook available at the visitor center points out geologic features along this trail.

The Dakota sandstone is about 30 million years younger than the Zuni sandstone, and was deposited on the eroded upper surface of the Zuni sandstone. The Dakota sandstone was deposited in water; its sand grains are not as well rounded and sorted as those of the wind-deposited Zuni sandstone, nor is its crossbedding as large in scale. It contains small pebbles which could not possibly have been carried by wind.

From the top of the mesa there is a good view of the racetrack valley of Triassic rocks extending northwest and southeast around the Zuni uplift. The Zuni Mountains are a slipper-shaped dome that exposes Precambrian granite at its core. Large parts of the core are still covered over with Permian rocks. Mesozoic rocks ring the uplift with hogbacks, cuestas, and intervening valleys like the Triassic racetrack, some of them partly filled in with lava flows.

## Gila Cliff Dwellings National Monument

Deep in the volcanic highlands of southwestern New Mexico, the Gila Cliff Dwellings are surrounded by coarse, pebble- and cobble-filled conglomerate. This rock, the Gila conglomerate, varies quite a bit from one place to another depending on the sources of its rock fragments. Here they are derived from volcanic ash and lava of the surrounding Tertiary volcanic range.

The cliff dwellings occupy alcoves in canyon walls of an uplifted

*The Gila Cliff Dwellings occupy large alcoves in massive ledges of Gila conglomerate.*

232

fault-block range that is part of the Basin and Range province. The region as a whole is volcanic; hot rocks below the surface still give up enough heat to supply nearby hot springs, and in places hot ground-water has leached overlying rock, causing scattered patches of soft white and yellow clay similar to that in Yellowstone Canyon in Yellowstone National Park—the rock that gave that park its name.

Corrugated upper canyon walls of Cliff Dweller Canyon are composed almost entirely of the Gila conglomerate. Exposure of this rock by stream erosion initiated development of the caves, which formed in weak zones where water seeping through porous rock layers gradually dissolved some of the calcium carbonate that cements the conglomerate together. With the help of wind and water, the caves deepened and grew larger, until about 700 years ago prehistoric peoples found shelter there.

## Salinas National Monument

Quarai, Abo, and Gran Quivira ruins, built by Pueblo Indians in the 12th to 14th centuries, lie in the Estancia Basin east of the Manzano Mountains. They and the churches added by 16th-century Spanish padres are built of local rock—Permian San Andres limestone at Gran Quivira, and red sandstone of the Permian Abo formation at Abo and Quarai. The early builders made full use of natural joints in these rocks, which allow them to be broken easily into more or less rectangular blocks.

The Estancia Basin held a lake in Pleistocene time. Beaches, bars, and spits can still be distinguished around the margins of the valley, particularly on aerial photographs. Ancient campsites on the old lakeshores show that hunter-gatherer peoples were in this area as early as 12,000 years ago. As the climate became drier after the close of the ice ages, Indians of the three pueblos cultivated corn, beans and squash—staple foods of that time—on the fine, loamy lake deposits.

Gran Quivira lies on a small mesa surfaced with San Andres limestone, a rock that hardened from limy muds made up of shells of

*The ruins at Abo are built of local Abo formation sandstone. The rock breaks naturally into thin, nearly rectangular blocks.*

*Red sandstone ledges near Abo line a shallow watercourse. After sudden storms, water plunging from these ledges excavates small plunge pools.*
National Park Service photo.

microscopic plants and animals that inhabited shallow Permian seas. The rock contains gypsum and salt, both of which flavor local water. Spanish historical documents tell that water for the village was obtained from rooftop runoff and from 32 walk-in wells, and stored in rock and clay cisterns. There are no nearby springs.

The pueblo of Quarai was more fortunate, obtaining water from several seeps. Close to the Manzano Range, a single east-tilted fault block, it was built where the Abo formation dips eastward off the range. This formation is 300 feet thick here and made up of many alternating layers of red shale and fine reddish brown sandstone, all deposited originally on a Permian river floodplain or delta. The rock contains some fossil plants and a few fossil vertebrate remains. Like the limestone at Gran Quivira, the rock is naturally squared by breaking along joints—an ideal building stone.

The pueblo of Abo is also close to the Manzano Mountains. Its ruins are in a valley between cuestas formed by the east-dipping Abo formation (west of the ruins) and the slightly younger Yeso formation

*Near Abo, steeply dipping red sandstone and shale of a Permian shore are exposed in a highway cut.*

234

*To forestall its own burial, a yucca growing on a dune face constantly lengthens its roots. Sand avalanches, some initiated by animal tracks, result from oversteepening of the leeward face.* National Park Service photo.

(east of the ruins). Hard sandstone units cap these cuestas, with softer shale and mudstone forming their slopes.

Not far from the three ruins, though not in the national monument, are a number of salt ponds and small playa lakes known as Laguna del Perro. Both Pueblo Indians and Spanish missionaries and settlers obtained salt there for domestic use and for trading. Some of the salt made its way to Mexico for use in smelting silver.

## White Sands National Monument

In the Tularosa Valley, between the San Andres and Sacramento mountains, three factors necessary for development of dunes come together: a source of sand, plenty of wind, and a place where the wind is forced to release whatever sand it has picked up. Here, the source is unusual: Though most dunes are composed of silica (quartz) sand, those at White Sands are composed of fine white grains of gypsum.

To create sand dunes, wind must blow 15 miles per hour or more— as it does here in February, March, and April. These spring winds blow from the southwest, where we must look for the sand source. A broad white playa, the bed of ancient Lake Otero, lies southwest of the present dunes in the deepest part of the Tularosa Valley. At its southern end is a smaller playa, Lake Lucero, which after heavy rains holds moisture even today. Much of the gypsum that ends up in the dunes comes from the floor of Lake Lucero.

*Some of the dune sand comes from dagger-like crystals of selenite, a form of gypsum, which occurs in lake sediments southwest of the White Sands.*

Groundwater beneath and around Lake Lucero is a source of more gypsum. In silty lake deposits that fringe the playa, capillary action draws gypsum-saturated ground water toward the surface, where it crystallizes as large, clear, daggerlike selenite crystals that seem to grow right out of the soil. These crystals, cracked and shattered by desert temperature changes, blasted by wind, break down into sand-sized particles that the wind can pick up and bounce across the lake flats to the dunes.

Within the dune field, wind is slowed down by turbulent flow over the dunes themselves. Wind bounces the sand grains up the long, gentle windward dune slopes. At dune crests, it loses its power and drops the grains. Accumulating near the dune crests, the sand eventually avalanches down the steep lee faces of the dunes.

The *ultimate* source of the gypsum is in Permian rocks of the San Andres Mountains—the mountains whose striped face marks the west side of the Tularosa Valley. It was deposited about 250 million years ago in an arm of a Permian sea, possibly an almost landlocked sea bordered, as the Red Sea is today, by deserts. At times, the sea water evaporated, leaving behind thick deposits of salt, gypsum, and other soluble minerals. The gypsum deposits alone may have been more than 500 feet thick.

The San Andres Mountains to the west and the Sacramento Mountains to the east developed late in Cenozoic time as a broad anticline arching across the future site of the Tularosa Valley. The valley came into existence in Miocene time, when tension in the Earth's crust, related to the formation of the Rio Grande Rift, caused the whole

wind

Knowledge of cross-bedding within a sand dune aids in identifying sandstone formed from ancient dunes. This drawing shows the inside of a 27-foot dune trenched for geologic studies. Since the prevailing wind is from the left, almost all laminae slope to the right.

center section of the great arch to collapse, to drop thousands of feet between bordering faults.

Collapse of the great arch exposed Permian rocks in both ranges—opening them to erosion and, in the case of the evaporite minerals, to solution. All material eroded from the two ranges ended up on the floor of Tularosa Valley, which has no outlet. Some of the gypsum dissolved from Permian rocks was redeposited in lake sediments of the closed basin, and some remained in the groundwater, later to recrystallize at the surface. From both lake deposits and surface crystals, particles of sand ride the wind to become part of the White Sands dune field.

Dunes have been classified by shape into a number of easily recognized types, four of which occur here. Where winds are strongest, on the southwest side of the dune field, low dome-shaped dunes occur. In the center of the dune field, where the sand supply is abundant but where winds are slightly weaker, crescent-shaped barchan dunes develop, along with long ridges of transverse dunes. At the margins of the dune field, where winds are not so strong, parabolic dunes form, their long trailing arms caught and anchored by desert vegetation.

The roots of shrubs hold sand in place as a dune moves on.

# Other Reading

Except for the excellent series, Scenic Trips to the Geologic Past, published by the New Mexico Bureau of Mines and Mineral Resources, books on New Mexico geology for the nonprofessional reader are few and far between. Some of those listed below pertain to specific areas within the state, or touch on geology only in passing. For readers who feel at home with geologic terminology, I recommend the field conference guidebooks prepared by the New Mexico Geological Society, the Four Corners Geological Society, and the West Texas Geological Society, available in most western university libraries as well as in other research libraries.

Atkinson, Richard, 1977. *White Sands: Wind, Sand and Time.* Southwest Parks and Monuments Association, Globe, Arizona.

Barnett, John, no date. *Carlsbad Caverns National Park.* Desert Press, Salt Lake City, Utah.

Chronic, Halka, 1986. *Pages of Stone, vol. 3: The Desert Southwest.* The Mountaineers, Seattle, Washington.

Colbert, Edwin Harris, 1983. *Dinosaurs of the Colorado Plateau.* Museum of Northern Arizona, Flagstaff, Arizona.

Kimbler, Frank S., and Narsavage, Robert J., Jr., 1981. *New Mexico Rocks and Minerals.* Sunstone Press, Santa Fe, New Mexico.

Ratkevich, Ron, and La Fon, Neal, 1978. *Field Guide to New Mexico Fossils.* Dinographic Southwest, Inc., Alamogordo, New Mexico.

Rigby, J. Keith, 1977. *Southern Colorado Plateau.* K/H Geology Field Guide Series, Kendall Hunt, Dubuque, Iowa.

Scenic Trips to the Geologic Past (various authors), New Mexico
Bureau of Mines and Mineral Resources, Socorro, New Mexico:
1. Santa Fe
2. Taos-Red River-Eagle Nest
3. Roswell-Ruidoso-Valley of Fires
4. Southern Zuni Mountains
5. Silver City-Santa Rita-Hurley
6. Trail Guide to Geology of the Upper Pecos
7. High Plains Northeastern New Mexico, Raton-Capulin
Mountain, Clayton
8. Mosaic of New Mexico's Scenery, Rocks, and History
9. Albuquerque—its Mountains, Valleys, Water, and
Volcanoes
10. Southwestern New Mexico
11. Cumbres and Toltec Scenic Railroad
12. The Story of Mining in New Mexico
13. Espanola-Chama-Taos
Trimble, Stephen, 1980. From Out of the Rocks:Discovering Ancient
Life in New Mexico. New Mexico State Life History Program.

# Glossary

**Agate:** a transparent, very finely crystalline form of quartz, usually with color bands or other patterns.

**Alluvial fan:** a sloping, fan-shaped mass of gravel and sand deposited by a stream as it issues from a narrow mountain canyon.

**Alluvial apron:** a skirtlike slope of gravel and sand formed by merging of alluvial fans; also called a bajada.

**Amphibole:** a group of dark, rodlike minerals, including hornblende; common in igneous rock.

**Amygdule:** a gas bubble or vesicle in volcanic rock that is filled with whitish minerals.

**Anhydrite:** a mineral similar to gypsum but containing no water.

**Anthracite:** hard coal.

**Anticline:** a fold that is convex upward.

**Aquifer:** a porous, permeable rock layer from which water may be obtained.

**Artesian:** water that rises above a water-bearing layer because of hydrostatic pressure.

**Ash, volcanic:** fine rock material thrown out by volcanoes.

**Ash fall:** volcanic ash settling from ash clouds that rise vertically from a volcano and then drift over wide areas.

**Ash flow:** hot, often incandescent volcanic ash buoyed up by its own expanding gases as it avalanches down a volcano's slope.

**Bajada:** a slope of gravel and sand formed by merging of alluvial fans.

**Basalt:** dark gray to black volcanic rock poor in silica and rich in iron and magnesium minerals.

**Beach ridge:** a ridge of sand and gravel thrown up by waves along a shore.

**Bed:** a single layer of sedimentary rock.

**Bedrock:** solid rock exposed at or near the surface.

**Bentonite:** soft, porous, light-colored rock formed by decomposition of volcanic ash.

**Blowout:** a shallow basin formed by wind erosion.

**Bomb, volcanic:** a boulder-sized blob of lava hurled from a volcano, usually rounded by its spin through the air.

240

**Borax:** an ore of boron, found in evaporite deposits of some alkaline or playa lakes.

**Boulder:** a rounded rock fragment with a diameter greater than 10 inches.

**Brachiopod:** a marine shellfish having two bilaterally symmetrical shells.

**Braided stream:** a stream that divides into an interlacing network of small channels.

**Breccia, volcanic:** volcanic rock consisting of broken rock fragments imbedded in finer material such as volcanic ash.

**Calcite:** a common rock-forming mineral, the principal mineral in limestone and travertine.

**Calcium carbonate:** calcite

**Caldera:** a basin-shaped, cliff-edged depression formed by a volcanic explosion and subsequent collapse of a volcano.

**Caliche:** a hard, crusty, whitish rock that accumulates near the surface as calcium carbonate and other minerals fill pore spaces in gravel.

**Cinder cone:** a small conical volcano formed of popcornlike volcanic material.

**Claystone:** sedimentary rock formed from clay.

**Cobble:** a rounded rock fragment with a diameter of 2.5 to 10 inches.

**Columnar jointing:** a pattern of vertical cooling joints creating many parallel columns in lava and volcanic ash.

**Conduit:** the feeder pipe of a volcano.

**Conglomerate:** rock composed of rounded, water-worn fragments of older rock, usually with coarse sand between the fragments.

**Continental rocks:** rocks making up the continents, lighter in color and density than those formed in ocean basins.

**Coral:** a group of marine animals that may deposit calcium carbonate in large reeflike masses.

**Crater:** the funnel-shaped hollow at the top of a volcano, from which volcanic material is ejected.

**Crinoid:** a group of marine animals having jointed stems and arms; often called sea lilies.

**Crust:** the outermost, cooled and hardened part of the Earth, averaging three miles thick under the oceans, 20 or 25 miles thick under the continents.

**Cuesta:** a ridge with a long, gentle slope formed by a resistant caprock, and a short steep slope on eroded edges of underlying rock.

**Desert pavement:** a veneer of tightly packed pebbles left when sand and silt are blown away by wind, common in arid climates.

**Desert varnish:** a dark, shiny surface of iron and manganese oxides, found on many exposed rock surfaces in desert regions.

**Dike:** a sheetlike intrusion of igneous rock resulting when magma intrudes and cools in a vertical crack or joint.

**Dip:** the direction and degree of tilt of sedimentary layers.

**Dome:** an anticline in which sedimentary rocks dip away in all directions.

**Dome, lava:** a rounded lava flow of very viscous (thick) lava that bulges upward directly over its volcanic vent.

**Dripstone:** travertine deposited by dripping water, as in stalactites and stalagmites.

**Evaporite:** a mineral deposited as mineralized water evaporates. Principal evaporite minerals are salt, gypsum, anhydrite, and potash.

**Extrusive rocks:** rock formed of magma which reaches the surface and solidifies there; also called volcanic rock.

**Fault:** a rock fracture along which displacement has occurred.

**Fault block:** a segment of the Earth's crust bounded on two or more sides by faults.

**Fault, listric:** a curved fault that is vertical at the surface but becomes more horizontal downward.

**Fault, normal:** a nearly vertical fault in which the overhanging side moves downward.

**Fault, reverse:** a nearly vertical fault in which the overhanging side moves upward.

**Fault scarp:** a steep slope or cliff formed by movement along a fault

**Fault, thrust:** a fault in which one side is pushed up and over the other side.

**Fault zone:** a zone of numerous small fractures that together make up many faults.

**Feldspar:** a group of common light-colored, rock-forming minerals containing aluminum oxides and silica.

**Floodplain:** relatively horizontal land adjacent to a river channel, with sand and gravel layers deposited by the river during floods.

**Flowstone:** travertine deposited in caves by water trickling across cave walls or floor.

**Fold:** a curve or bend in rock strata.

**Formation:** a mappable unit of stratified rock.

**Fossil:** remains or traces of a plant or animal preserved in rock.

**Gabbro:** a dark gray to black, crystalline igneous rock, the intrusive equivalent of basalt.

**Geode:** a hollow or partly hollow, roughly spherical rock containing minerals deposited from solution in groundwater.

**Geothermal:** pertaining to the Earth's heat.

**Gneiss:** banded metamorphic rock thought to form from granite (which it commonly resembles) or sandstone.

**Granite:** a coarse-grained igneous intrusive rock composed of chunky crystals of quartz and feldspar peppered with dark biotite and hornblende.

**Gravel:** a mixture of pebble, boulders, and sand not yet consolidated into rock.

**Greenstone:** greenish metamorphic rock thought to form from volcanic rocks.

**Groundwater:** subsurface water, as distinct from rivers, streams, seas, and lakes.

**Group:** a major unit of stratified rock composed of several related formations.

**Gypsum:** a common mineral, calcium sulfate, formed usually by evaporation of sea water.

**Hematite:** a common iron oxide mineral; also an ore of iron.

**Hogback:** a long, narrow ridge with a sharp crest formed by erosion of layers of resistant rock tilted approximately 45 degrees.

**Hydrostatic pressure:** the pressure exerted by confined water due to the weight of water at higher levels.

**Igneous rock:** rock formed from molten magma.

**Intrusion:** a body of igneous rock intruded into older rock while molten.

**Intrusive rock:** igneous rock created by subsurface cooling of molten magma.

**Joint:** a rock fracture along which no significant movement has taken place.

**Laccolith:** a lenslike intrusion that spreads between rock layers, doming those above it.

**Lava:** igneous rock created by cooling of molten magma on the Earth's surface or under the sea.

**Lava dome:** a dome-shaped body of volcanic rock formed from very thick magma.

**Lava tube:** a tunnellike cave left where very fluid lava flowed out from under its own cooling crust.

**Limestone:** a sedimentary rock consisting of calcium carbonate, usually formed from the shells of marine animals and calcium-secreting plants.

**Limonite:** a yellow-brown iron oxide mineral.

**Listric fault:** see Fault, listric.

**Lithology:** the physical character of rocks.

**Lithosphere:** the solid outer portion of the Earth, consisting of the crust and upper mantle.

**Lode gold:** gold occurring in veins in the rock.

**Magma:** molten rock.

**Magma chamber:** a reservoir of magma from which volcanic materials are derived.

**Magnetic reversal:** a shift in the Earth's magnetic field, with the positive (north) and negative (south) magnetic poles switching places.

**Mantle:** the thick zone between the Earth's core and crust.

**Meander:** a bend in a river.

**Mesa:** a flat-topped hill or mountain capped with a resistant rock layer and edged with steep cliffs or slopes.

**Metamorphic rocks:** rocks formed from older rocks by great heat and pressure or by chemical changes.

**Metamorphism:** the alteration of rocks by great heat and pressure or by chemical change.

**Mica:** a group of minerals characterized by their separation into thin, shiny plates or flakes.

**Mid-ocean ridge:** a volcanic ridge running across ocean basins, the site of sea floor spreading.

**Mineral:** a naturally occurring substance with a characteristic chemical composition and usually with typical color, texture, and crystal form.

**Monocline:** a fold in stratified rock in which all the strata dip in the same direction.

**Mud crack:** a shrinkage crack in drying mud.

**Mudflow:** a flowing mass of mud.

**Mudstone:** sedimentary rock formed from mud.

**Normal fault:** see Fault, normal.

**Oceanic rocks:** rocks making up the oceanic crust, usually darker and denser than those of the continents.

**Olivine:** an olive green or yellow mineral, common in basalt magma whose source is in the Earth's mantle.

**Parabolic dune:** a horseshoe-shaped sand dune in which long arms partly stabilized by vegetation point upwind.

**Pebble:** a rock fragment, commonly rounded, 0.2 to 2.5 inches in diameter.

**Pediment:** a gently inclined erosion surface carved in bedrock at the base of a mountain range.

**Pegmatite:** very coarse-grained igneous rock similar to granite, occurring in veins or irregular masses in intrusive igneous rock.

**Period:** a subdivision of geologic time shorter than an era, longer than an epoch.

**Placer gold:** gold found as free particles or flakes in gravel and sand.

**Plate:** a large block of the Earth's crust, separated from other blocks by mid-ocean ridges, trenches, and collision zones.

**Plateau:** a flat-topped mountain larger than a mesa, usually capped with a resistant rock layer and edged on two or more sides with cliffs or steep slopes.

**Playa:** a flat-floored dry lake bottom in an undrained desert basin.

**Playa lake:** a shallow intermittent lake formed on a playa after rains.

**Polarity reversal:** the switching of the Earth's magnetic poles from positive (now north) to negative (now south) or negative to positive.

**Porphyry:** an igneous rock containing conspicuous large crystals in a fine-grained matrix.

**Potash:** potassium carbonate used in the manufacture of fertilizer.

**Pressure ridge:** a ridge on a lava flow buckled up by pressure of flowing material.

**Pumice:** light-colored, frothy volcanic rock, often light enough to float on water.

**Quartz:** a hard, glassy, rock-forming mineral composed of crystalline silica

**Racetrack valley:** a circular or oval valley formed by erosion of soft rock layers around a dome of harder rock.

**Radiometric dating:** dating of rocks by measurement of radioactive minerals and their decay products.

**Reverse fault:** see Fault, reverse.

**Ripple marks:** ripples on rock surfaces formed from ripples in the sand or silt of which they were formed.

**Rift:** a narrow block of the Earth's crust formed by downdropping between two more or less parallel fault zones that reach down to the mantle.

**Rift valley:** a) a valley formed by a river following a pre-existing rift; b) a valley formed by subsidence of the Earth's crust and caused by large-scale extension of the crust.

244

**Rock glacier:** a glacierlike tongue of broken rock, usually lubricated by water and ice and moving slowly like a true glacier.

**Salt:** sodium chloride; the term may be used to include other evaporite minerals as well.

**Sandstone:** sedimentary rock formed from sand, usually cemented with calcium carbonate.

**Scarp:** a low cliff caused by movement along a fault.

**Schist:** metamorphic rock whose parallel orientation of mica flakes causes it to break easily along parallel planes.

**Sedimentary rock:** rock formed from particles of other rock transported and deposited by water, wind, or ice.

**Selenite:** a silky, fibrous variety of gypsum common in veins.

**Shale:** a fine-grained sedimentary rock formed by consolidation of clay, silt, or mud, characterized by breaking into flat sheets. The term is also loosely applied to include siltstone, mudstone, and claystone that do not break in this fashion.

**Sheet flood:** an expanse of moving water that spreads thinly over a large area without concentrating in channels.

**Silicic:** containing more than 65% of silica in silicate minerals.

**Siltstone:** rock formed from particles smaller than those of sand and larger than those of clay.

**Sill:** a thin body of igneous rock intruded between horizontal rock layers.

**Sink:** a depression caused when ground collapses into an underlying solution cavern.

**Spatter cone:** a small, steep-sided cone built by molten lava spattering from a volcanic vent.

**Squeeze-up:** thick lava squeezed upward through a crack in a lava crust.

**Stalactite:** a dripstone "icicle" hanging from the roof of a limestone cavern.

**Stalagmite:** a dripstone pedestal rising from the floor of a limestone cavern.

**Strata:** layers of sedimentary rock. Singular is stratum.

**Stratified:** formed in layers, as sedimentary rock.

**Stratigraphic:** pertaining to layered or stratified rocks.

**Stratovolcano:** a cone-shaped volcano built of alternating layers of lava and volcanic ash.

**Syncline:** a troughlike downward fold in sedimentary rocks.

**Tar sand:** sand or sandstone saturated with tar, a very thick oil.

**Terrace:** a relatively level bench bordering a river valley; generally representing an earlier river floodplain.

**Time scale:** a geologic calendar dividing the history of the Earth into units of time.

**Trap:** a structure in rock layers that tends to capture oil below an impermeable layer.

**Transverse dunes:** a sand dune in the form of a long, often sinuous ridge at right angles to the prevailing wind direction.

**Travertine:** hot spring deposits composed largely of calcite.

**Trench:** a long, deep oceanic depression formed when an oceanic plate of the Earth's crust is pulled downward under the edge of an overriding continental plate.

**Tuff:** a rock formed of compacted volcanic ash and cinders.

**Tuff, ashfall:** a rock formed from volcanic ash falling from an overhead ash cloud.

**Tuff, ashflow:** a rock formed from very hot volcanic ash that flows down the side of a volcano.

**Tuff, welded:** see Tuff, ashflow.

**Unconformity:** a surface of erosion or non-deposition that separates younger strata from older rocks.

**Valley fill:** sand, gravel, and volcanic material filling a valley.

**Varve:** a very thin layer of sediment deposited in a quiet lake in one year's time.

**Vent:** an opening through which volcanic material escapes to the Earth's surface.

**Vesicle:** a bubble cavity in lava.

**Volcanic glass:** lava that has congealed without formation of crystals.

**Volcanic neck:** an erosional remnant of volcanic rock that formerly filled a volcano's conduit.

**Volcanic rock:** rock formed by cooling of molten magma on or very near the surface.

**Volcano:** a mountain or smaller structure built by escape of magma to the Earth's surface.

**Volcano, composite:** see Stratovolcano.

**Volcano, shield:** a dome-shaped volcano formed by moderately fluid lava.

**Water table:** the surface below which rocks and soil are saturated with groundwater.

**Wildcat well:** an oil well drilled without definite indications that oil is present.

246

# Index

254

Check for our books at your local bookstore. Most stores will be happy to order any which they do not stock. We encourage you to patronize your local bookstore. Or order directly from us, either by mail, using the enclosed order form or our toll-free number, 1-800-234-5308, and putting your order on your Mastercard or Visa charge card. We will gladly send you a complete catalog upon request.

## Some other geology titles of interest:

| | |
|---|---|
| ____ROADSIDE GEOLOGY OF ALASKA | 14.00 |
| ____ROADSIDE GEOLOGY OF ARIZONA | 15.00 |
| ____ROADSIDE GEOLOGY OF COLORADO | 15.00 |
| ____ROADSIDE GEOLOGY OF IDAHO | 15.00 |
| ____ROADSIDE GEOLOGY OF MONTANA | 15.00 |
| ____ROADSIDE GEOLOGY OF NEW MEXICO | 14.00 |
| ____ROADSIDE GEOLOGY OF NEW YORK | 14.00 |
| ____ROADSIDE GEOLOGY OF NORTHERN CALIFORNIA | 14.00 |
| ____ROADSIDE GEOLOGY OF OREGON | 14.00 |
| ____ROADSIDE GEOLOGY OF PENNSYLVANIA | 15.00 |
| ____ROADSIDE GEOLOGY OF TEXAS | 16.00 |
| ____ROADSIDE GEOLOGY OF UTAH | 14.00 |
| ____ROADSIDE GEOLOGY OF VERMONT & NEW HAMPSHIRE | 10.00 |
| ____ROADSIDE GEOLOGY OF VIRGINIA | 12.00 |
| ____ROADSIDE GEOLOGY OF WASHINGTON | 15.00 |
| ____ROADSIDE GEOLOGY OF WYOMING | 12.00 |
| ____ROADSIDE GEOLOGY OF THE YELLOWSTONE COUNTRY | 10.00 |
| ____AGENTS OF CHAOS | 12.95 |
| ____FIRE MOUNTAINS OF THE WEST | 16.00 |
| ____GEOLOGY UNDERFOOT IN SOUTHERN CALIFORNIA | 12.00 |

Please include $2.00 per order to cover postage and handling.

Please send the books marked above. I enclosed $_____

Name_____

Address_____

City _____ State _____ Zip _____

☐ Payment Enclosed (check or money order in U.S. funds)
Bill my: ☐VISA ☐ MasterCard   Expiration Date: _____

Card No. _____

Signature _____

## MOUNTAIN PRESS PUBLISHING COMPANY
### P.O. Box 2399 • Missoula, MT 59806
### ☎ Order Toll-Free 1-800-234-5308 ☎
### *Have your MasterCard or Visa ready.*